U.S.-Japan Technology Linkages In Biotechnology: Challenges for the 1990s

*A15045 541171

Committee on Japan
Office of Japan Affairs

Office of International Affairs
National Research Council

National Academy Press
Washington, D.C. 1992

NOTICE: The project that is the subject of this report was approved by the Governing Board of the National Research Council, whose members are drawn from the councils of the National Academy of Sciences, the National Academy of Engineering, and the Institute of Medicine. The members of the committee responsible for the report were chosen for their special competences and with regard for appropriate balance.

This report has been reviewed by a group other than the authors according to procedures approved by a Report Review Committee consisting of members of the National Academy of Sciences, the National Academy of Engineering, and the Institute of Medicine.

The National Academy of Sciences is a private, nonprofit self-perpetuating society of distinguished scholars engaged in scientific and engineering research, dedicated to the furtherance of science and technology and to their use for the general welfare. Upon the authority of the charter granted to it by the Congress in 1863, the Academy has a mandate that requires it to advise the federal government on scientific and technical matters. Dr. Frank Press is president of the National Academy of Sciences.

The National Academy of Engineering was established in 1964, under the charter of the National Academy of Sciences, as a parallel organization of outstanding engineers. It is autonomous in its administration and in the selection of its members, sharing with the National Academy of Sciences the responsibility for advising the federal government. The National Academy of Engineering also sponsors engineering programs aimed at meeting national needs, encourages education and research, and recognizes the superior achievement of engineers. Dr. Robert M. White is president of the National Academy of Engineering.

The Institute of Medicine was established in 1970 by the National Academy of Sciences to secure the services of eminent members of appropriate professions in the examination of policy matters pertaining to the health of the public. The Institute acts under the responsibility given to the National Academy of Sciences by its congressional charter to be an adviser to the federal government and, upon its own initiative, to identify issues of medical care, research, and education. Dr. Kenneth I. Shine is president of the Institute of Medicine.

The National Research Council was organized by the National Academy of Sciences in 1916 to associate the broad community of science and technology with the Academy's purposes of furthering knowledge and advising the federal government. Functioning in accordance with general policies determined by the Academy, the Council has become the principal operating agency of both the National Academy of Sciences and the National Academy of Engineering in providing services to the government, the public, and the scientific and engineering communities. The Council is administered jointly by both Academies and the Institute of Medicine. Dr. Frank Press and Dr. Robert M. White are chairman and vice-chairman, respectively, of the National Research Council.

This report was prepared with support of a grant from the United States-Japan Foundation. Available from:

Office of Japan Affairs
National Research Council
2101 Constitution Avenue, N.W.
Washington, DC 20418

National Academy Press
2101 Constitution Ave., N.W.
Washington, DC 20418

Library of Congress Catalog Card Number 92-60203
International Standard Book Number 0-309-04699-8
S568

First Printing, May 1992
Second Printing, June 1992
Third Printing, September 1992

Copyright ∀ 1992 by the National Academy of Sciences

Printed in the United States of America

COMMITTEE ON JAPAN

Harold Brown (*Chairman*)
Johns Hopkins Foreign Policy Institute

Erich Bloch (*Vice-Chairman*)
Council on Competitiveness

C. Fred Bergsten
Institute for International
  Economics

Lewis M. Branscomb
Harvard University

Lawrence W. Clarkson
The Boeing Co.

I. M. Destler
University of Maryland

Mildred S. Dresselhaus
Massachusetts Institute of
  Technology

Daniel J. Fink
D. J. Fink Associates, Inc.

Ellen L. Frost
Institute for International
  Economics

Lester C. Krogh
3M Co.

E. Floyd Kvamme
Kleiner Perkins Caufield & Byers

Yoshio Nishi
Hewlett-Packard Co.

Daniel I. Okimoto
Stanford University

John D. Rockefeller IV
United States Senate

Richard J. Samuels
MIT Japan Program

Robert A. Scalapino
University of California, Berkeley

Hubert J. P. Schoemaker
Centocor, Inc.

Ora E. Smith
Illinois Superconductor Corp.

Albert D. Wheelon
Hughes Aircraft Co. (retired)

Ex Officio Members:

Gerald P. Dinneen, Foreign Secretary, National Academy of Engineering

James B. Wyngaarden, Foreign Secretary, National Academy of Sciences
  and Institute of Medicine

BIOTECHNOLOGY WORKING GROUP ON PRIVATE SECTOR TECHNOLOGICAL LINKS BETWEEN THE UNITED STATES AND JAPAN

Hubert J. P. Schoemaker (*Co-Chairman*)
Centocor, Inc.

G. Steven Burrill (*Co-Chairman*)
Ernst & Young

Mark D. Dibner
North Carolina Biotechnology Center

Stelios Papadopoulos
PaineWebber

James B. Wyngaarden
National Research Council

Robert T. Yuan
University of Maryland

OFFICE OF JAPAN AFFAIRS

Since 1985 the National Academy of Sciences and the National Academy of Engineering have engaged in a series of high-level discussions on advanced technology and the international environment with a counterpart group of Japanese scientists, engineers, and industrialists. One outcome of these discussions was a deepened understanding of the importance of promoting a more balanced two-way flow of people and information between the research and development systems in the two countries. Another result was a broader recognition of the need to address the science and technology policy issues increasingly central to a changing U.S.-Japan relationship. In 1987 the National Research Council, the operating arm of both the National Academy of Sciences and the National Academy of Engineering, authorized first-year funding for a new Office of Japan Affairs (OJA). This newest program element of the Office of International Affairs was formally established in the spring of 1988.

The primary objectives of OJA are to provide a resource to the Academy complex and the broader U.S. science and engineering communities for information on Japanese science and technology, to promote better working relationships between the technical communities in the two countries by developing a process of deepened dialogue on issues of mutual concern, and to address policy issues surrounding a changing U.S.-Japan science and technology relationship.

Staff

Martha Caldwell Harris, Director
Thomas Arrison, Research Associate
Maki Fife, Program Assistant

Contents

1. INTRODUCTION 1

2. TECHNOLOGY LINKAGES_DEFINITIONS AND
   APPROACHES TO ANALYSIS 5

3. TECHNOLOGY LINKAGES_SCOPE, SIGNIFICANCE
   AND TRENDS 13
   The Actors, 13
   Company-to-Company Linkages Between the
      United States and Japan, 20
   Corporate Strategies in the United States and Japan, 25
   Special Characteristics of Japanese Investment in the
      U.S. Biotechnology Industry, 29
   Technology Linkages Between Japanese Companies and
      U.S. Universities and Nonprofit Research Institutions, 33
   Examples of Technology Linkages_Multiple Purposes
      and Mechanisms, 41

4. PROSPECTS FOR THE FUTURE 45

5. CONCLUSIONS 51

APPENDIXES

A. Case Studies of U.S.-Japan Technology
   Linkages in Biotechnology 61
   Case I: Calgene-Kirin, 61
   Case II: Monotech, Inc. and Showa-Toyo Diagnostics, 67
   Case III: Kirin-Amgen, 74
   Case IV: Hitachi Chemical Research-University of
      California, Irvine, 81

B. Examples of Linkages Between Japanese Companies and U.S.
   Academic Research Institutions 90

C. Workshop on U.S.-Japan Technology Linkages in Biotechnology
   Agenda and Participants 97

1

Introduction

The prevailing view is that the United States is a world leader in biotechnology.[1] U.S. researchers excel in basic science, and U.S. industry has moved new ideas to the market by commercializing technology. In fiscal year 1990 alone, the federal government provided more than $3.5 billion in funding for biotechnology R&D and U.S. industry invested approximately $2 billion.[2] Approximately 50 to 75 biotechnology companies were formed

---

[1]See, for example, Office of Technology Assessment (OTA), *Biotechnology in a Global Economy* (Washington, D.C.: U.S. Government Printing Office, October 1991), p. 19, and *New Developments in Biotechnology* (Washington, D.C.: U.S. Government Printing Office, 1988), p. 3. In the OTA report, biotechnology is broadly defined to include any technique that uses living organisms (or parts of organisms) to make or modify products, to improve plants or animals, or to develop microorganisms for specific use. For general statements on the state of the U.S. biotechnology industry, see also Japanese Technology Evaluation Center (JTEC), "JTEC Panel Report on Biotechnology," June 1985; Ministry of International Trade and Industry (MITI), *Sangyo Gijutsu no Doko to Kadai* (Trends and Topics in Industrial Technology) (Tokyo: Tsushosangyosho, 1988); George B. Rathmann, "An Industry View of the Public Policy Issues in the Development of Biotechnology," in John R. Fowler III, ed., *Application of Biotechnology: Environmental and Policy Issues* (Boulder: Westview Press, 1987). See also Mark D. Dibner and R. Steven White, "Biotechnology in the United States and Japan: Who's First?" *Biopharm*, March 1989.

[2]The President's Council on Competitiveness, "Report on National Biotechnology Policy," p. 6. G. Steven Burrill and Kenneth B. Lee, Jr. estimate that in 1991, the federal government invested $3.7 billion and industry $3.2 billion in biotechnology-related R&D. See G. Steven Burrill and Kenneth B. Lee, Jr., *Biotech 92: Promise to Reality* (San Francisco, Ca.: Ernst & Young, 1991).

each year during the decade of the 1980s, over 1,000 in the last 20 years. The biotechnology and pharmaceutical industries have been rated as second only to the computer software and services sector in terms of total value creation among U.S. high-technology companies founded since 1965.[3]

Is this rosy view of U.S. preeminence_across the board from basic to applied biotechnology R&D, to commercialization and global market competitiveness_accurate and will it persist? [4] Another, perhaps better, way to pose the question is to ask whether the United States will remain competitive and reap a "fair share" of future profits from the significant investments made in biotechnology. These broad questions set the context for this report, which assesses technology linkages between the United States and Japan. The purpose of this study is not only to examine the scope and nature of technology linkages between the United States and Japan but also to consider the forces behind these linkages as well as the future impact on competitiveness for the organizations involved and for the United States as a country.

To summarize some of the major themes, the study suggests that there are a number of powerful forces driving an expansion of technological linkages of many types between the United States and Japan. We are moving toward a global economy, and the desires of large Japanese companies, both pharmaceutical companies and ones doing business in unrelated fields, to access technology developed in the United States and to compete globally are important contributing factors. Japanese firms see biotechnology as a way to use scarce resources to improve their productivity and international competitiveness. For nonpharmaceutical companies, biotechnology is a technological tool allowing diversification into new, higher value-added product areas. From the U.S. perspective, a driving force for small innovative biotechnology firms is the need for capital to fuel their R&D, thus stimulating relationships of various kinds with large capital-rich Japanese companies. Another stimulus is the desire of large U.S. pharmaceutical companies and biotechnology firms to access the Japanese market.

Increased cooperation between the United States and Japan is desirable and inevitable as biotechnology becomes part of an increasingly global economy and technology base. In this context of increasing cooperation, the question is whether the U.S. biotechnology industry will continue to compete effectively. To do so, it will be necessary to structure technology linkages with Japan to ensure that U.S. participants gain clear benefits.

This study documents a prevailing pattern of transfer of biotechnology developed in the United States to Japan during the past two decades. The analysis in this report suggests that the linkages formed so far serve as

---

[3]See Arthur D. Little and HOLT Value Associates, "The Upside 100," *Upside*, December 1990, p. 25. Value creation was measured in a number of ways, including shareholder value, for each firm since its establishment.

[4]For a more sober view, see President's Council on Competitiveness, op. cit.

mechanisms primarily for technology transfer from the United States to Japan. Looking at past patterns, some wonder whether the technology has been sold too cheaply and whether U.S. firms can develop effective strategies for making technology linkages with Japan work to their advantage in the future.

There are new trends, such as the establishment of Japanese "offshore" R&D facilities in the United States and growing investments by Japan in basic research, that hold a potential for learning from Japan. Increasingly, Japanese companies are building ties to American universities through training, research grants, and endowed chairs. Japan's strength in areas such as bioprocessing technologies suggests potential areas for future technology transfer from Japan to U.S. biotechnology firms.

From the perspective of individual U.S. biotechnology firms or larger companies, it may be possible or even necessary to ensure corporate growth (and possibly survival) by linking up with Japanese companies in joint ventures or other agreements that give the Japanese partners rights to license and market technologies and products that were developed in the United States. Over time, however, the result may be to create a significant competitive challenge in both the U.S. market and global competition unless these alliances are developed such that the U.S. firms benefit through the development of improved manufacturing and marketing capabilities.

The implications for the United States as a country must also be considered. Researchers from around the world are drawn to the open and excellent biotechnology research laboratories of U.S. universities and research institutions_organizations financed with taxpayer funds. Being first in basic science, however, in no way ensures that U.S. companies will compete effectively at home and around the world. Japan, a country where the primary emphasis has been on technology commercialization, benefits greatly from access to fundamental research carried out in the United States. Given the considerable investments that the United States has made in supporting biotechnology R&D, it may be appropriate to consider new policy approaches that ensure that the United States maintains its lead in global competition. Government and industry in Japan have identified biotechnology as a key technology for future industrial growth and are working together to increase R&D investments in this field.[5] Should we do likewise?

---

[5]Estimates of expenditures by the government of Japan for biotechnology-related R&D vary, for reasons that will be outlined in detail later. According to research by the NRC working group, in 1991 expenditures increased approximately 16 percent over the previous year to a total of 89.6 billion yen. According to estimates by the U.S. Department of State (unclassified cable of July 1990), Japanese public and private spending on rDNA totaled 57 billion yen, and total R&D on biotechnology-related work for all Japan Bioindustry Association (JBA) members totaled 276 billion yen. See *Heisei Yonnendo Kaku Shocho Baiteku Kanren Yosan Seifu Genan* (Japan Fiscal Year 1992 Biotech-Related Budged Proposal), in *Biosaiensu to Indasutori* (Bioscience and Industry), March 1992, pp. 277-285. See Table 2 for more detail.

This report was prepared by a working group of experts, as part of a project initiated by the National Research Council's Committee on Japan to examine technology linkages between Japan and the United States. Co-chaired by Hubert Schoemaker of Centocor and G. Steven Burrill of Ernst & Young, the working group was formed in the fall of 1990 and met a number of times in 1991 to deliberate and confer on the data collection process. A workshop on U.S.-Japan Technology Linkages in Biotechnology was convened in June 1991 to gain additional insights from other experts in the United States and Japan. The staff of the National Research Council's Office of Japan Affairs, which also serves as the staff for the Committee on Japan, assisted the working group in data collection, and analysis and compilation of results.

2

Technology Linkages—Definitions and Approaches to Analysis

Biotechnology is a research- and capital-intensive industry for which intellectual property rights protection and government regulation are critically important. The industry is growing rapidly, both domestically and internationally, and the context is rapidly changing.[6] Linkages between U.S. and foreign-based biotechnology companies also are expanding, but there is no consensus about the long-term impacts. Will Chugai's acquisition of a majority interest in Gen-Probe or Roche's acquisition of Genentech lead to the creation of potent competing firms, or will these linkages bring new strength to U.S. industry and the U.S. economy? Will Hitachi's investment in an R&D laboratory on a University of California campus bring benefits to both sides? Put another way, will biotechnology go the way of the semiconductor industry to face severe competition from Japanese companies that focus their efforts on commercialization of technology that originated here?

This report was compiled to assess the nature, scope, and impacts of technology linkages between the United States and Japan in biotechnology and to outline policy issues for government, industry, and universities. The major focus is on commercial biotechnology—the use of biotechnological tools to develop and manufacture products for the market. The line be-

---

[6]See, for reference, Burrill and Lee, *Biotech '92*, op. cit.; G. Steven Burrill and Kenneth B. Lee, Jr., *Biotech 91: A Changing Environment* (San Francisco, Ca.: Ernst & Young, 1990); and Biotechnology Information Division, North Carolina Biotechnology Center, *Biotechnology in the U.S. Pharmaceutical Industry* (Research Triangle Park, N.C.: NCBC, 1990).

tween basic research and commercial biotechnology is not hard and fast, however. Companies focusing their efforts on the commercialization of biotechnology are research intensive, carefully watching the work going on in basic research laboratories because new developments in science can become the basis for new products seemingly overnight. But bringing these products to market can take a number of years, particularly in the health care field. Erythropoietin (EPO), for example, generated $200 million in revenues for Amgen in its first full year of sales in 1990. Amgen carried out research to bring this product to the clinical trial stage for approximately 3 years, and it took another 3 years to complete clinical trials and obtain regulatory approval before going to market.[7] Because of the importance of fundamental research to firms seeking to commercialize biotechnology, the working group decided to include in its analysis linkages formed between Japanese firms and research laboratories at U.S. universities, national laboratories, and biotechnology centers that are likely to have an impact on market competition.

Biotechnology is a diverse activity comprised of many scientific disciplines. Indeed, some prefer not to call it an industry because developments in biotechnology research span many fields of science and affect a wide range of industries (see Figure 1). For the purposes of this report, the working group has defined biotechnology as any activity, product, or process that involves recombinant DNA and/or cell fusion technology. These technologies are currently applied to develop products for human health care, specialty chemicals and biosensors, and human and agricultural applications and to improve the generation of energy and protection of the environment. More than 100 large chemical, pharmaceutical, and agricultural companies use biological processes. Large pharmaceutical and agricultural firms are using biotechnological techniques to complement their established in-house research efforts. These large companies should be distinguished from the dedicated biotechnology firms (many of them small firms formed by some of our nation's premier researchers and entrepreneurs) that focus almost exclusively on the use of biotechnology to develop new products through biological processes. In terms of market segments, health care (including human diagnostics, vaccines, and therapeutics) is by far the largest.[8]

---

[7]See Gary P. Pisano, "Joint Ventures and Collaboration in the Biotechnology Industry," David C. Mowery, ed., *International Collaborative Ventures in U.S. Manufacturing* (Washington, D.C.: American Enterprise Institute, 1988), p. 199 for an estimate that the development of a pharmaceutical product takes 5 to 10 years from the initiation of basic research to marketing of the product.

[8]There are more than 1,000 biotechnology companies in the United States, about 76 percent of them small companies with 1 to 50 employees. (See Burrill and Lee, *Biotech 91*, op. cit., pp. 15-16.) The *Biotechnology in Japan Yearbook 1990/91* states that there are more than 800 Japanese companies involved in biotechnology commercialization and estimates the 1990

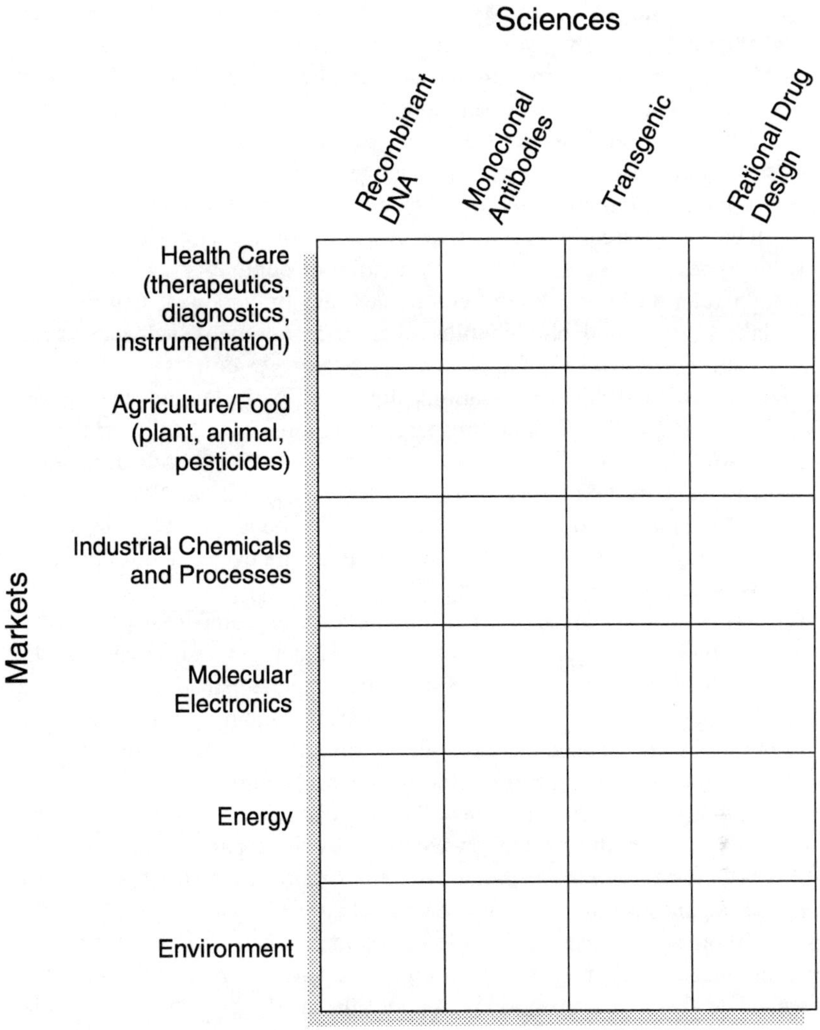

FIGURE 1 Matrix definition of biotechnology.

Sciences (columns): Recombinant DNA, Monoclonal Antibodies, Transgenic, Rational Drug Design

Markets (rows): Health Care (therapeutics, diagnostics, instrumentation); Agriculture/Food (plant, animal, pesticides); Industrial Chemicals and Processes; Molecular Electronics; Energy; Environment

---

market in Japan for biotechnology-related products as more than 100 billion yen. Note that the Japanese count includes companies that are involved in biotechnology in some way; a large number of these companies have their primary business in some other area. See Mark D. Dibner and R.S. White, *Biotechnology Guide USA* (London: MacMillan, 1991), for a list of 742 biotechnology firms and 142 corporations involved in biotechnology in the United States.

Defining the term "technology linkages" is equally complex. Linkages include company-to-company activities such as marketing, sales, distribution and/or manufacturing, inward and outward licensing of technology, and various types of equity investments and R&D collaborations. As will be discussed in more detail in the following section, technology linkages between companies are the most prominent and most studied types_both domestically and internationally_but the degree of actual technology transfer involved varies greatly and must be evaluated on a case-by-case basis.

Other types of linkages relevant to a study of commercial biotechnology include relationships between companies and universities, national research laboratories, and biotechnology centers. In many instances these research laboratories are supported in part by taxpayer dollars. Companies establish ties with these organizations not only by endowing chairs and providing grants for facilities and research programs, but also by establishing links with individual professors through contract research and other mechanisms such as laboratory visits and training of employees. Conferences and specialized journals also offer mechanisms for learning about new developments in biotechnology R&D, as do patent registrations, cell line deposits, and related documents.

While the primary focus of attention has been on company-to-company linkages in biotechnology, consider the following hypothetical case as an example of how universities can be important mechanisms. A researcher from a U.S. university is invited to give a research seminar at another U.S. institution, unaware that the biotechnology program at the host institution is generously funded by a company based in Japan. Details from the presentation are quickly faxed to the firm's Tokyo headquarters, where they are used as the basis for filing patent applications by the Japanese company. In Japan, where the principle for patent rights is first to file rather than first to invent, the Japanese company stands a good chance of securing patent rights. Consider another example that illustrates the importance of scientific publication in one country to research around the globe. A young Japanese researcher, Masashi Yanagisawa, read about the work of Highsmith and his colleagues on cell membrane receptors for a family of peptides called endothelins. The young researcher persuaded his professor that this was a worthy topic for a Ph.D. dissertation, and a group of researchers at Tsukuba University began work that led to a breakthrough published in *Nature* in March 1988. Two independent groups in Japan continue path-breaking work in this area, while Japanese pharmaceutical companies race to find potential therapeutic agents.[9]

---

[9]See John Vane, "Endothelins Come Home to Roost," *Nature*, vol. 348, December 20-27, 1990, p. 673.

Linkages provide opportunities not only for a transfer of technology and products but also for access to capital, market, and distribution channels; improved manufacturing capability; regulatory expertise; and research strengths. The creation or transfer of technology, whether consciously intended or an indirect result, is a prerequisite for a "technology linkage." One can study technology linkages by combing the trade press and other specialized publications for reports of specific interactions or deals between individual companies. This will provide a representative but not a complete accounting of either relationships among companies or the biotechnology-related in-house efforts of large pharmaceutical and other companies.[10] In many cases linkages between U.S. and Japanese organizations are complex and encompass a variety of mechanisms that evolve over time.

It is also important to underscore the ambiguity that arises in defining a "U.S." or a "foreign" firm. For years the standard approach has been to use equity ownership as the criterion for making the distinction. In practice, U.S. policy has been "national treatment" for foreign investors in the United States and the reduction of foreign barriers to investment overseas. Foreign investment has played a critical role in U.S. economic development, and U.S. multinational companies have grown through investments overseas, particularly in Europe where restrictions have been less extensive than those of Japan before the 1980s.

For the purposes of this study, the critical elements in distinguishing between foreign and domestic firms are the location of a firm's headquarters (or where most of its employees are working) and majority ownership by citizens of a country. This definition is practical but not entirely satisfactory from an analytical perspective. About one-fourth of U.S. biotechnology firms are publicly owned, but companies based in Japan and other countries are often privately owned, and the details of ownership and control are less accessible. Nor should U.S. ownership be equated with U.S. interests.[11] A "foreign" firm that operates manufacturing and R&D facilities in the United States may, under certain conditions, contribute more significantly to the U.S. economic and technology base than a "U.S." firm that moves its manufacturing and R&D overseas. Realities such as these complicate analysis of technology linkages and must be kept in mind.

To assess technology linkages between the United States and Japan in biotechnology, the working group developed a multidimensional matrix (see Figure 2). Linkage mechanisms, organizations involved, and industries make

---

[10]Readers should note that the focus of this report is on the linkages *among* firms rather than the internal biotechnology-related efforts of larger firms. Readers interested in the internal activities of larger firms can consult other studies, such as OTA, *Biotechnology in a Global Economy*, op. cit.

[11]Robert Reich, "Who is Us?" *Harvard Business Review*, January-February, 1990, pp. 53-64.

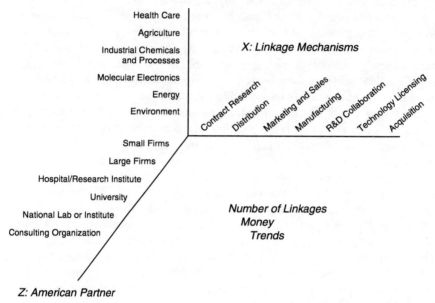

FIGURE 2 Matrix of U.S.-Japan private sector biotechnology linkages.

up the three axes of the matrix. As will be discussed in more detail later, the predominant pattern during the past decade has been linkages involving emerging U.S. firms in the health care field in technology licensing and product marketing agreements with large Japanese firms. The analysis necessarily focuses on this area where U.S.-Japan linkages in biotechnology have been formed, but also includes treatment of potential future linkages in Chapter 4.

Finally, a few preliminary words about underlying assumptions and goals are in order. Many studies have focused on international competition in biotechnology. This study focuses on technology linkages and long-term impacts, for the following reasons. First, Japan is now a technological superpower, and the United States and Japan must develop new modes of interacting that involve reciprocal transfers of Japanese technology and manufacturing expertise and/or commensurate contributions to basic research, the source of much of the technology. While the standard assumption is that Japan lags behind the United States in biotechnology, a study by the U.S. Department of Commerce identified biotechnology as an emerging technology where the United States may fall behind Japan in the years ahead.[12] At the very least, Japan has technological strengths in certain areas, such as bioprocessing and biosensors, worthy of attention. In addition, Japanese industry reports increasing investments in basic research,[13] and the Japanese government is acting to catalyze new efforts in generic technology development. Nor should Japan's strengths in manufacturing and global marketing be underestimated.[14] For all these reasons, a basic assumption is that Japan represents a major competitive challenge in biotechnology.

Second, technology linkages present opportunities for two-way flows of technology. Press accounts provide clues to the direction of technology transfer and the terms of the arrangements, but more detailed knowledge often is required to evaluate the impact. As the competitive challenge to U.S. industry has grown in recent years, increased attention has been paid to past patterns of technology licensing to Japanese firms and training of Japanese scientists and engineers in fields such as electronics. An examination of technology linkages should contribute to an understanding of the expectations and results, and how linkage mechanisms can be structured so that both sides benefit. A basic assumption underlying this study is that these linkages are a fact of life, and it is important for U.S. companies and the United States as a nation to develop more effective strategies to ensure that benefits flow to both sides.

A good deal of attention has been paid to structural differences in industrial organization and markets in Japan and the United States. Japan's biotechnology industry contrasts with the U.S. biotechnology industry in several ways. In Japan the primary players are large integrated pharmaceutical firms joined by large firms seeking to diversify into new businesses. There are also differences in the regulatory environments, drug pricing, and medical practices that relate to cultural differences between the two countries. One such difference lies in the fact that Japanese physicians prescribe and dispense drugs and profit directly from sales.

Only a small number of the Japanese companies active in the U.S. market are biotechnology based, and many new drugs reflect joint development with a U.S. partner, but large Japanese companies are playing an increasingly important role in the U.S. market. Japanese pharmaceutical

---

[12] U.S. Department of Commerce, *Emerging Technologies: A Survey of Technical and Economic Opportunities* (Washington, D.C.: Department of Commerce, 1990).

[13] The "basic research" under way in Japanese corporations is *mokuteki kiso kenkyu* (translated goal-oriented basic research) and would not be recognized as such by many people in the United States.

[14] Dibner and White, op. cit.

firms are developing new drugs, marketing them in the United States, and in some cases licensing them to U.S. companies.[15]

In the face of such striking differences in industry organization, patent systems, regulation, and medical practices in the two countries, observers point to a "playing field" that is not level.[16] The National Research Council's (NRC) working group discussed this issue at some length and concluded that the purpose of this assessment is not to develop recommendations that will create a "level playing field." In view of the significant differences and asymmetries in funding and access to research, technology, and markets, it seems correct to assume that the differences will not be eliminated quickly. Instead of trying to create a level playing field, the more important question is how to compete and win in this context.

Increasing U.S.-Japan technology linkages are part of a global phenomenon. Linkages are affected by capital markets, the macroeconomic environment, scientific prowess, patent systems, and other factors that vary across countries and regions. One interesting question is whether the U.S.-Japan linkages are different or generally similar to linkages between U.S. and European firms. Fortunately, the research carried out by members of the NRC working group and others on U.S.-European linkages can be drawn on to set the context and form contrasts and comparisons.[17]

There is no guarantee that the future will repeat recent experience_that the United States will maintain a competitive edge. New factors that may influence the future development of biotechnology as a global enterprise must be taken into account if the United States is to maintain its position.

---

[15] Cardizem and Cefobid are among the biggest-selling drugs in the U.S. market. Cardizem, sold by Marion Merrell Dow, was licensed from Tanabe of Japan. Cefobid is marketed by Pfizer under license from Toyama Chemical Company of Japan.

[16] Mark D. Dibner, "Drug Regulation in Japan: Can We Compete on Their Playing Field?" *Biopharm*, vol. 2, no. 9, 1989, pp. 34-42.

[17] Lois Peters, in a study that focused on relationships between Japanese and European pharmaceutical industries, found evidence of technology transfer from Japan to Europe, particularly through the establishment of laboratories in Japan by European companies. See Lois Peters, "Emerging Private Sector Alliances," in Herbert I. Fusfeld, ed., *Changing Global Patterns of Research and Development* (Rochester, N.Y.: Center for Science and Technology Policy, Rensselaer Polytechnic Institute, 1990). Note that the data were not disaggregated with an analysis of interactions in biotechnology.

3

Technology Linkages—Scope, Significance, and Trends

THE ACTORS

To understand why technology linkages are being formed, it is important to consider the special characteristics of the major actors—small U.S. biotechnology firms, large established U.S. companies, the federal and state governments in the United States, U.S. universities and research institutions, large Japanese companies, and Japanese government agencies that provide funding for biotechnology R&D. The major impetus for the formation of technological linkages is the development and exploitation of biotechnology. Each of these actors brings different resources to bear in linkages that take many different forms.

Small U.S. biotechnology firms (sometimes called "dedicated" or "new" biotechnology companies) are those formed for the sole purpose of commercializing biotechnology. The formation of these small firms was spurred by the development of recombinant DNA and monoclonal antibody technologies in the 1970s. These technologies, which emerged from universities and national research institutes, were public and widely diffused, stimulating the formation of new biotechnology firms by venture capitalists in association with entrepreneurs and university research scientists. In 1981, a peak year for the formation of biotechnology firms, almost 70 new companies were established.[18]

---

[18] See OTA, *New Developments in Biotechnology*, op. cit., p. 79.

Such small biotechnology firms continue to generate much of the most promising research in biotechnology. One recent study concluded that small biotechnology firms make unusual contributions to innovation, as measured in patent applications (both product and process). Although they no longer have the overwhelming innovative advantage vis _ vis established U.S. or Japanese companies seen in the early 1980s, the patents they spawn are still cited disproportionately.[19] R&D is the lifeblood of biotechnology firms for many U.S. firms whose R&D expenditures well exceed revenues (see Table 1).

In contrast to the small U.S. biotechnology firms that have consistently had a comparative advantage in biotechnology R&D, most large U.S. companies did not have their own in-house biotechnology R&D programs in the early 1980s.[20] Instead, they relied on R&D contracted with the small biotechnology firms. Japanese and European companies also lacked in-house biotechnology R&D programs in the early 1980s. Today, some of the large pharmaceutical and other companies are beginning to pursue biotechnology-related R&D_as a complement, rather than a substitute, to their main areas of business activity.

The special strengths of the large companies continue to be in traditional drug discovery, manufacturing, marketing and distribution of products, and their financial strength. In addition, large pharmaceutical firms have much experience with the process of regulatory approval, which can be time consuming and costly in the United States and elsewhere. Large pharmaceutical companies invest considerable resources in drug discovery and development as a prerequisite for manufacturing and marketing.

The large U.S. firms such as Monsanto, Eli Lilly, Schering-Plough, and Merck, which were the first to begin their own in-house biotechnology R&D programs in the early 1980s, also established technology links with the small biotechnology firms. The major motivation for these linkages was to access technology developed in the small biotechnology firms in order to commercialize it and to bring the technology in-house over time.[21]

Universities and other research institutions continue to be critical actors in biotechnology research. As noted earlier, basic research in biochemistry and molecular biology at universities can lead directly to commercial applications. For example, Centocor pays royalties to universities for 15 products it has developed. Moreover, individual researchers trained at universi-

---

[19] Joshua Lerner, "The Flow of Intellectual Property Between the U.S. and Japanese Biotechnology Industries," Harvard Business School Working Paper, 1991. Lerner's paper summarizes his detailed analysis of patent citations as a measurement of technological flows. His conclusions confirm the predominant flow of technology from the United States to Japan.

[20] Office of Technology Assessment, *Commercial Biotechnology* (Washington, D.C.: U.S. Government Printing Office, 1984).

[21] Mark D. Dibner, "Corporate Strategies for Involvement in Biotechnology," *Bifutur* (Paris), July-August, 1987, pp. 47-48.

TABLE 1 Top 10 U.S. Biotechnology Firms in R&D Spending, 1990

	FY 1990 (millions)	Revenues (millions)
Genentech	$173	$447
Amgen	63	190
Genetics Institute	61	40
Cetus	56	39
Chiron	50	79
Centocor	46	65
Biogen	36	50
Xoma	28	20
Immunex	19	31
Genzyme	19	50

NOTE: R&D and revenue figures have been rounded to nearest million.

SOURCE: PaineWebber, Inc., December 1991.

ties become not only scientific leaders but also entrepreneurial leaders in the new biotechnology firms. In biotechnology more than in perhaps any other industry, companies see linkages to universities as a fast track to new ideas. There are hundreds of collaborative arrangements between biotechnology companies and U.S. universities (and nonprofit research institutions), many focused on human pharmaceutical applications.

The U.S. government plays a powerful role in the development of biotechnology. Its two principal activities are in research and the regulation of new biotechnology products. The National Institutes of Health (NIH) represents one of the largest biomedical research complexes in the world. NIH and, to a lesser degree, the National Science Foundation (NSF) fund most of the basic biological research at universities and nonprofit research institutes in the United States. Regulatory functions are split among a number of agencies: the U.S. Food and Drug Administration (drugs, food); the U.S. Environmental Protection Agency (environmental regulations); the U.S. Department of Agriculture (plants, animals); and NIH (research guidelines). Although progress has been made toward development of a unified regulatory scheme, there has been considerable criticism of the slow approval process for new products, particularly in medical products. Such delays can only increase the financial burden for the companies involved. For large companies, drug pricing is a major issue.[22] Unlike the Japanese government,

---

[22] Alice Sapienza has argued that, while the federal government is encouraging "hi or biotech" advanced drugs in the health care arena, pressures on the pharmaceutical industry in the form of drug pricing (health care cost containment) are creating a situation where the products will not be paid for. See "Irreconcilable Differences? Strategic Innovations for a Publicly Insured Market" *Technovation*, forthcoming.

the U.S. government has played a limited role in technology development and transfer. Creation of the BioProcessing Center at the Massachusetts Institute of Technology (MIT) and passage of the Technology Transfer and Orphan Drug acts represent infrequent examples of government action that may help speed up the commercialization of research.

State governments have become increasingly involved in biotechnology even though their expenditures are minimal compared to those of the federal government. Their principal effort has been in the creation of state biotechnology centers, many of which carry out basic research in areas that might be relevant to the states' economies. Some states, such as Maryland, Massachusetts, Pennsylvania, and North Carolina, have begun to experiment with new approaches to commercialization, with fostering the creation of new biotechnology companies, and with the promotion of sales of biotechnology products overseas.

In Japan there are virtually no U.S.-style small biotechnology companies. A variety of possible explanations for this can be offered, but the lack of a dynamic venture capital industry, a centralized R&D process in large traditional Japanese firms, and the comparative lack of movement of professionals from company to company are certainly important factors. Japanese companies active in biotechnology are mostly large, well-established pharmaceutical, fermentation, or chemical companies, such as Yamanouchi, Kirin, and Mitsubishi Kasei. In recent years other Japanese companies (even steel and tobacco companies) have entered the biotechnology industry in order to diversify into new businesses.[23] In a recent survey of 1,600 CEOs, R&D directors, and business planners in Japan's largest companies, biotechnology was selected as the most important technology for the future.[24] Fumio Kodama notes Japan's high expectations for biotechnology in the 5- to 10-year time period.[25]

The Japanese government is another important actor in the promotion of biotechnology, although Japanese government funding in all areas of R&D (biotechnology included) is dwarfed by the investments made by companies.[26] The Japanese government, particularly the Ministry of International Trade and Industry (MITI), nevertheless played a significant role in

---

[23] Toyota is beginning biotechnology research in its corporate laboratory, according to a December 20, 1990 report of the *Nihon Keizai Shimbun* (in Japanese), p. 11. See also report by Robert K. Fujimura for U.S. Department of Commerce, *R&D in Biotechnology-Related Industries in Japan*, 1989, NTIS PB 89-167936.

[24] "Bio, Kankyo nado Juyo ni" (Importance of Bio, Environment, etc.), *Nihon Keizai Shimbun*, September 9, 1990, p. 6.

[25] Fumio Kodama, comments at Workshop on U.S.-Japan Technology Linkages in Biotechnology, June 12, 1991.

[26] See Robert T. Yuan and Mark D. Dibner, *Japanese Biotechnology: A Comprehensive Study of Government Policy, R&D and Industry* (London: MacMillan, 1990) and Mark D. Dibner and R. Steven White, *Biotechnology Japan* (New York: McGraw-Hill, 1989).

stimulating interest in biotechnology in the late 1970s and early 1980s, leading a variety of companies to establish internal goals in this field or to join R&D collaborations with other companies.

As Table 2 shows, the overall amount of money spent on biotechnology through the general account (government of Japan budget) is less than one-fifth of that spent by the U.S. government, but it is important to look closer to get an accurate picture of Japanese government support.[27]

In contrast to the emphasis on support for basic research in the United States, the share of Japanese government funding for university research has declined as a part of the national R&D effort in recent years. Still, the Ministry of Education reports that 40 percent of the grants to university researchers under the *kagaku kenkyu hi* (scientific research fund) go to life sciences and that many of the priority areas selected for preferential treatment in the awards process are in biotechnology.[28] At the same time, MITI, the Science and Technology Agency, and other ministries are increasing their funding of biotechnology-related R&D, including $27 million in 1991 for international collaboration in the Human Frontier Science Program.

Table 3 shows that the share of Japanese contributions in life sciences in leading journals has remained steady in recent years. (Japan's overall contribution, however, remains about one-fifth that of the United States.[29])

Perhaps the most striking aspect of Japanese government support for biotechnology is its commercial orientation and the number of agencies involved. One example that illustrates both of these aspects of policy is support for research on the role of carbohydrates in cell function. Three Japanese agencies launched projects in this field in March 1991, bringing the total number of agencies involved to five. Three of these projects encompass participation by industry.[30] Other institutes in protein engineering and marine sciences also have been formed over the past few years

---

[27]The statistics here are taken from a Japanese trade journal, *Baiosaiensu to Indasutori* (Bioscience and Industry), January 1990 and February 1991. Estimating the Japanese government's biotechnology budget presents several problems. First, the definition of biotechnology used by the Japanese government appears to be broader than that used in the United States, which would introduce an upward bias in the figures. Second, the figures do not include loans made through the Japan Development Bank, loans through the Small and Medium Size Business Program, and sizable tax breaks on R&D and the purchase of laboratory equipment.

[28]"Heisei 2 Nendo Kakushocho Baiteku Kanren Yosanan" (1990 Biotechnology-Related Budget Proposal by Agency), *Baiosaiensu to Indasutori*, January 1990, p. 97.

[29]National Science Board, *Science and Engineering Indicators, 1991*, p. 388. Between 1981 and 1987, Japan's contributions increased from 6.2 percent to 7.1 percent of the world total of publications in biomedical research, while the U.S. share declined slightly from 39.5 percent to 38.2 percent.

[30] "San Shocho de Shin Purojekuto" (Three Agencies Launch Projects), *Nihon Keizai Shimbun*, September 9, 1990, p. 17. Genzyme Japan will receive approximately $1.7 million over the next 5 years from the MITI to conduct carbohydrate research; foreign firms can participate.

TABLE 2 Japanese Government's Biotechnology-Related Budget

	1988	1989	1990	1991	1992 (requested)
Ministry of International Trade and Industry (MITI)					
General Account (billion yen)	5.1	7.1	6.6	9.9	10.3
(million $)	37.8	52.6	48.6	73.4	76.4
(Investment Account million $)	(20.7)	(15.6)	(NA)	(NA)	(NA)
Science and Technology Agency (STA)					
General Account (billion yen)	13.8	18.2	17.6	20.3	22.6
(million $)	102.2	134.8	130.3	150.3	167.7
(Loan Account million $)	(50.4)	(65.9)	(66.7)	(44.7)	(8.0)
Ministry of Education (Mombusho)					
Program Funding (billion yen)	14.0	14.5	16.1	19.8	19.3
(million $)	103.7	107.4	119.3	146.7	143.0
40% of Research Subsidies					
(billion yen)	19.6	21.0	22.3	23.6	25.8
(million $)	145.2	155.6	165.3	174.8	191.4
Total General Account					
(billion yen)	33.6	35.5	38.4	43.4	45.1
(million $)	248.9	263.0	284.4	321.5	334.4
Ministry of Health and Welfare					
General Account (billion yen)	4.8	6.0	6.6	7.4	8.5*
(million $)	35.6	44.4	48.9	55.0	63.1
(Investment Account million $)	(19.3)	(17.8)	(17.0)	(17.0)	(17.0)
Environment Agency					
General Account (billion yen)	0.34	0.34	0.3	0.34	0.45*
(million $)	2.5	2.5	2.2	2.5	3.4
Ministry of Agriculture, Forestry, and Fisheries					
General Account (billion yen)	6.6	7.5	7.9	8.3	9.1
(million $)	48.9	55.6	58.5	61.7	67.7
Total General Account (billion yen)	64.2	74.6	77.4	89.6	96.0
(million $)	475.6	522.6	573.3	664.3	712.7
Change in General Account	(NA)	+16%	+4%	+16%	+7%
(Financing Accounts million $)	(90.4)	(99.3)	(NA)	(NA)	(NA)

NOTE: Conversions at 135 yen per dollar. Items that have an impact on biotechnology but that do not appear in the budget include private sector funding for university research administered by the Ministry of Education, extramural support for Ministry of Health and Welfare research institutes, loans extended through the Japan Development Bank and the Small- and Medium-Sized Business Program, biotechnology-oriented ERATO programs administered by STA, and R&D subsidies given as tax breaks.

*In 1992, the Ministry of Health and Welfare and the Environment Agency changed their definitions of biotechnology, making them more inclusive. The figures here are based on the definition used in previous budgets.

SOURCE: Compiled by OJA Staff from figures appearing in *Baiosaiensu to Indasutori* (Bioscience and Industry), January 1990, February 1991 and March 1992; and figures provided by the Ministry of International Trade and Industry.

TABLE 3 Japanese Papers Published in Leading Journals

	Japanese Papers Published as a Percentage of Total Papers	
	1980-1984	1985-1989
Biology		
Journal of Biological Chemistry	5.1	7.2
EMBO Journal	2.2	4.1
Biochemical Journal	2.9	3.5
Molecular and Cellular Biology	2.0	2.5
Cell	1.6	1.6
Average	3.9	5.2
Multidisciplinary		
Nature	1.6	1.8
Science	0.7	0.8
Average	1.3	1.4

SOURCE: Institute for Scientific Information, *Science Citation Index*, 1980-1989, as related in "Japanese Scientists Increase Their Presence in World-Class Journals," *Science Watch*, May 1990, p. 7. In the article, John Tooze, editor of *EMBO Journal*, noted that the Japanese are strongest in biochemistry and fields relevant to the pharmaceuticals industry. He also notes that the Japanese share of papers rose only modestly in *Science, Nature,* and *Cell,* "the top three journals in biology."

under the Key Technology Center program, which features strong industry leadership. The U.S. government could actively participate with industry in the development and exploitation of commercial applications of biotechnology, as discussed in more detail in the conclusions chapter of this report.

In the 1980s Japanese companies began to build competitive strategies featuring expanded participation in the U.S. research community and market. One indicator is the fact that they have filed many pharmaceutical patents in the United States. These patents are cited often, but they are less science intensive than the U.S.-origin patents filed at the same time.[31] In terms of new nonbiotechnology drugs introduced into the market, the growing contributions of large Japanese firms are clear. In biotechnology, Japanese companies gradually built strength during the 1980s by perfecting manufacturing technology through automation and other means in areas such as bioprocessing, by commercializing technology and products licensed from U.S. companies, and by deepening their independent R&D capabilities. Suntory

---

[31] Francis Narin and Dominic Olivastro, *Identifying Areas of Leading Edge Japanese Science and Technology*, CHI Research for NSF, April 15, 1988.

is building a completely automated factory for the production of biotechnology-based drugs.[32]

Generalizations about the six major groups of actors must be qualified in light of changes now under way. There is evidence to support the thesis that a few U.S. biotechnology firms are today moving toward "forward integration," establishing their own manufacturing, marketing, and sales capabilities (rather than relying on the large firms to manufacture and sell the products they develop or joining them in joint ventures). Forward integration, however, may not be easy for even the most successful biotechnology firms. There is also some evidence that the large U.S. companies are moving to expand their in-house biotechnology R&D. Meanwhile, Japanese companies are expanding their ties to innovative U.S. firms and increasing R&D in more fundamental research areas. A distinguishing characteristic of large Japanese firms, particularly pharmaceutical firms, seems to be their interest in using biotechnology as the driving force in their attempt to become serious global players, rather than as a complement to established business activities.

COMPANY-TO-COMPANY LINKAGES BETWEEN THE UNITED STATES AND JAPAN

Few studies exist that focus explicitly on technology linkages between U.S. and Japanese firms or that document changes over time.[33] Since the mid-1980s, however, it has been clear that linkages between U.S. biotechnology firms and foreign companies have been expanding and that linkages with Japanese firms have been significant. The NRC biotechnology working group assembled data on linkages from a number of sources, including the data base developed by the North Carolina Biotechnology Center, Bioscan, reports by Ernst & Young, JETRO, JEI, and other proprietary sources. Together, these sources provide an overview of the various linkage mechanisms. In many instances, however, information in specialized journals and data bases must be augmented with expert knowledge to draw conclusions about the direction of technology transfer and the significance for corporate strategy.

There are many ways to classify technology linkages, but an important distinction can be made, at least in theory, between those that involve the commercialization of technology already in existence and those established

---

[32] "Santori_Baio iyaku no Zenjido Kojo" (Suntory_A Completely Automated Factory to Produce Bio Drugs), *Nihon Keizai Shimbun*, September 26, 1991.

[33] In addition to the works of Mark D. Dibner cited previously, see Lois S. Peters, *Technical Network Between U.S. and Japanese Industry* (Rochester, N.Y.: Center for Science and Technology Policy, Rensselaer Polytechnic Institute, 1987), p. 117 ff. See also Donald H. Dalton and Phyllis A. Genther, U.S. Department of Commerce, *The Role of Corporate Linkages in U.S.-Japan Technology Transfer 1991*, NTIS PB 91-165571, 1991.

with the purpose of developing new technology. Licensing and marketing agreements, materials supply, and some types of joint ventures not oriented to new technology development are formed to exploit technology already brought through development and manufacturing. Research contracts (which usually include licensing agreements), joint development agreements to produce a new product or process, and equity investments oriented around the development of new technology are examples of linkages aimed at developing new technology. Generally speaking, the first type of technology linkage (designed to transfer established technology) requires less certainty and less tacit knowledge about a particular partner and its R&D process than do technology linkages for the development of new technology, where equity investments are also more common.[34]

Data collected by the NRC working group show that during the decade of 1981 to 1991 the most common form of technology linkage between U.S. and Japanese firms was of the first type_ a transfer of technology developed in the United States to a Japanese company through a licensing or marketing agreement. About half of the linkages included in the data base involved licensing of rights to manufacture a product (23.8 percent) or licensing of marketing rights (27.3 percent) to a Japanese company (see Table 4). Research contracts, direct acquisitions, and equity investments (for a minority stake in a U.S. company) have been much less prominent.

It is important to emphasize, however, that most alliances are multifaceted. They frequently include a technology license, an R&D collaboration, some marketing, manufacturing and distribution rights, and in some cases an equity investment. Trade press and other published reports, the basis for data compilation, typically report on some new development and often do not include a complete review of all aspects of technology linkage, including those that are ended.

The predominant pattern for U.S.-Japan linkages in biotechnology is a tie-up between a small U.S. biotechnology company and a large Japanese company (see Table 5). Overall, 200 of the 282 cases in the data base involved a linkage between a small U.S. biotechnology firm and a large Japanese company. The overwhelming majority of these linkages (160) were in the health care field; of the remaining 40 cases, more than half were in agriculture and food-related technologies. Table 5 provides a summary of the linkage patterns. Of the 51 linkages between large U.S. companies and large Japanese ones, it should be noted that more than half involve technology transfers in areas other than biotechnology precisely defined (such as traditional pharmaceuticals). While these data make it clear that

---

[34] See Gary P. Pisano, "Using Equity Participation to Support Exchange: Evidence from the Biotechnology Industry," *Journal of Law, Economics and Organization*, vol. 5, no. 1, Spring 1989.

TABLE 4 Alliances Between U.S. Biotechnology Firms and Large Japanese Companies, 1981-1991

Type of Alliance	Comments	% of Total
Acquisition	Outright purchase of a company	2.1
Equity purchase	Purchase of a minority stake in a company	8.2
Joint deal	Unspecified alliance, usually for product development	21.6
Research contract	Biotech firm is paid for R&D on a specified product or product line	4.6
Joint venture	New joint venture company formed	9.6
Licensing agreement	License for rights to a product or technology, often for a limited geographic region	23.8
Marketing agreement	License to market a product or technology	27.3
Purchase of material or service	Provision of biological materials, products, or services for a fee	2.8

NOTE: The table was compiled from 282 cases that involve alliance formation. The data include 27 cases involving large U.S. companies active in biotechnology in which the focus of the alliance was not primarily biotechnology, and 12 alliances in biotechnology equipment.

SOURCE: North Carolina Biotechnology Center, Institute of Biotechnology Information, Actions Database.

technology linkages established to date have focused on the health care sector, it is important to remember that linkages in biotechnology applications in agriculture, the environment, and bioelectronics will probably increase in the future.

Although the numbers are small, there is some evidence of an increase in equity investments in recent years. A trend toward increasing numbers of marketing agreements, a type of relationship in which there is often limited technology transfer, is quite clear.

With regard to the direction of technology flow, there is no question that the predominant pattern of technology transfer has been and remains from the United States to Japan. In more than 90 percent of the linkages between small U.S. firms and large Japanese companies where the direction of technology flow could be established, it was from the United States to Japan. When the 231 cases involving small U.S. firms were reviewed, this pattern persisted. In only 11 cases was there clear evidence of technology

TABLE 5 U.S.-Japan Corporate Technology Links in Biotechnology, 1981-1991

Category	Total	Cases Where Technology Flow Is Identified	% Flow to Japan
Small U.S. firm Large Japanese firm Health care	160	154	90
		There appears to be an increase in the importance of marketing agreements over time, but the proportion of linkages in which technology flows to Japan has remained constant.	
Small U.S. firm Large Japanese firm Nonhealth care	40	31	90
		About half are targeted at the agriculture/food markets. There are no apparent trends in industries, mechanisms, or technology flow. The percentage of linkages in which technology flows to Japan is equivalent to that of health care.	
Small U.S. firm Small Japanese firm All markets	31	25	96
		The first two cases of small Japanese companies purchasing equity in U.S. firms occurred in 1990.	

NOTE: Table includes 10 alliances in biotechnology equipment. Table does not include 24 cases involving large U.S. companies and large Japanese companies in biotechnology. Based on expert review of the 24 case sample, 10 involved technology transfer from the United States to Japan and 4 involved transfer from Japan to the United States; in 10 cases it was not possible to determine the direction of technology transfer. It should be noted that three of the four cases involving technology transfer to the United States were transactions making up the Upjohn-Chugai partnership in which the original technology was developed in the United States by Genetics Institute and licensed to Chugai. Considering the small number of cases involving large U.S. companies and the large weight that would be assigned to the Upjohn-Chugai partnership, the NRC working group decided that inclusion of these cases in the table would be misleading.

SOURCE: North Carolina Biotechnology Center, Institute of Biotechnology Information, Actions Database.

transfer from Japan to the United States. In only another eight cases was there clear evidence of a two-way flow of technology.

A recent analysis by Weijan Shan and William Hamilton confirms these trends. Shan, using BioScan data and a very detailed disaggregation of linkage types, found that the majority of U.S.-Japan cases involved technology transfer rather than joint development of new biotechnology products. Shan takes this as evidence that U.S. firms avoid joint development and manufacturing relationships that may provide access to new technology but are much more willing to transfer technology when they feel more confident in valuing it and protecting intellectual property rights.[35]

It is important to note that the statistics provide no basis for judging whether the U.S. organization attempted to negotiate an arrangement to acquire technology from the Japanese company. Nor do the data provide a basis to judge technology transfers that often occur at a later stage, after the companies have worked together for some time. The Kirin-Amgen relationship, discussed later, falls into that category. Transfers of process technology and engineering skills are more subtle and difficult to measure than product transfers. These types of transfer occur at a later stage in a relationship and are not necessarily captured in published reports on corporate linkages.

The evidence on technology flows also reflects, to some extent, a life-cycle phenomenon. Biotechnology is a new field, and most of the action is in the area of product development where U.S. firms are strong. Japan's strengths in bioprocessing suggest areas for future cooperation and possible technology transfer from Japan as more products are developed. While the United States also possesses strengths in bioprocessing, this may be an opportunity for "reverse flow" of technology from Japan in the future.

Technology linkages between U.S. and Japanese companies increased sharply in 1987 and 1988, probably due to capital market constraints in the United States during that period. During the past three years, U.S.-Japan linkages have decreased in number, due perhaps to some disappointment in Japan that investments in biotechnology will not have near-term payoffs.[36]

In general, there have been three times as many linkages between Japanese companies and U.S. companies than among Japanese companies or between Japanese companies and firms in Europe or other parts of Asia. Japan clearly continues to look to the United States for products and technology. At the same time, there have been many more linkages *among* U.S. firms than between U.S. and Japanese firms.[37]

In joint ventures and joint development projects, there may be opportunities for technology transfer from Japan to the United States. In only a few unusual instances such as the Kirin-Amgen joint venture, however, is there clear evidence of a two-way flow. Kirin reportedly made an important

---

[35] Weijan Shan and William Hamilton, "Country-Specific Advantage and International Cooperation," *Strategic Management Journal*, vol. 12, 1991, pp. 419-432.

[36] Isao Karube of Tokyo University noted the disappointment in Japan in the Worshop on U.S.-Japan Technology Linkages in Biotechnology, June 12, 1991.

[37] The clear majority (62.6 percent) of the cases in the North Carolina Biotechnology Center data base for the 1985-1989 period were U.S./U.S. cases; U.S. linkages with Japan made up 17 percent of the total and U.S. linkages with Europe 11 percent. Data presented at the NRC Workshop on U.S.-Japan Technology Linkages in Biotechnology, June 12, 1991.

contribution to Amgen's manufacturing technology.[38] In most cases the transfer of technology from Japan has been in traditional pharmaceutical rather than biotechnology-based products. Additional avenues for technology transfer may be opened as Japanese firms deepen their capabilities in fundamental research, as they perfect manufacturing processes through automation and other means, and if U.S. firms negotiate linkages that feature a transfer of manufacturing or other biotechnology know-how from Japan.

The *Yano Report*, a Japanese publication, provides another set of data points. The report shows that during the 1987-1989 period Japanese *pharmaceutical* firms increasingly obtained products from new U.S. partners. This report indicates an upswing in codevelopment ventures with non-Japanese firms and a definite decline in cross-licensing. During the 3-year period, new licensing from foreign firms and codevelopment with non-Japanese firms made up 22 and 38 percent, respectively, of the total 226 cases.[39] It may be that Japanese pharmaceutical companies are becoming more interested in codevelopment with foreign partners, but there is no way to determine whether the major activity is in biotechnology.

How can the pattern of continuing technology transfers from the United States to Japan be explained? Will it continue in the 1990s, and what are the implications for the United States? To answer the first question, it is important to understand the strategies of corporate leaders in the United States and Japan. The other two questions will be addressed in later sections of this report.

CORPORATE STRATEGIES IN THE UNITED STATES AND JAPAN

Because the overwhelming majority of U.S.-Japan biotechnology linkages are in the health care field, the working group focused on corporate strategies in this area. The perspectives of CEOs in small U.S. biotechnology companies, and in large U.S. or Japanese companies using biotechnology for health care, provide sharp contrasts that help to explain the pattern of technology linkages noted above (see Table 6).

Consider, first, the importance of biotechnology to various types of companies. For the small U.S. biotechnology firm (SBF), biotechnology is the reason for existence, and the focus of corporate strategy is on new technology development. For large U.S. companies (LUCs) using biotech-

---

[38] Amgen uses roller bottles to swish nutrients over the gene-spliced cells that produce EPO. Kirin provided an automated roller bottle handling machine. See "Can Amgen Follow Its Own Tough Act?" *Business Week*, March 11, 1991, p. 95. For more details, see Case Study III in Appendix A of this report.

[39] See *SCRIP*, No. 1516, May 23, 1990, p. 25.

TABLE 6 CEO Perspectives on the Role of Biotechnology in Health Care

Important Factors	Company		
	Small Biotechnology Firm (SBF)	Large U.S. Company (LUC)	Large Japanese Company (LJC)
Biotechnology importance	A Justifies existence and technical focus	C Enabling technology and toolbox; some new products	B New products critically important
Importance of company-company linkages	A *Must* make alliances with Japanese companies to have access to capital and Japanese market	C Develop own Japanese subsidiary with R&D, manufacturing, and marketing strategy to strengthen distribution	B *Must* introduce newly developed product and enter European and U.S. markets
Global presence	C Objective is to become self-financing and market products first in U.S. and then Europe; financial constraint to introduce new products in Japan	A Global orientation critical for long-term competition and financial return	B Global strategy to enter U.S. and Europe through alliances
Domestic government role	A Very important for patents, approval, pricing, and "help" for innovators	C Important for pricing and tax considerations	B Introduction of new technology; protect Japanese market through regulated processes, pricing, patents
Academe (U.S. and Japan)	A Products produced in U.S. academe critically important; little interaction with Japanese universities	C Provide researchers, test own ideas, do R&D in-house, NIH, beginning interactions with Japanese universities	B Cherry pick in U.S. universities and monitor access to Japanese universities and R&D
Capital	A A must	B Earnings per share pressure	C Long-term perspective

continued

TABLE 6 *Continued*

Important Factors	Company		
	Small Biotechnology Firm (SBF)	Large U.S. Company (LUC)	Large Japanese Company (LJC)
Competitive environment	A Survival	B Independence or merger	B Global growth
Japanese market	C	B Diversification and growth	A Critical
U.S. market	A A must	B Defensive	B Growth and new product development
Future role year 2000+	A Less than 5% will truly succeed	B Consolidation will continue; critical mass name of the game	C Enter U.S. and Europe through joint ventures

NOTE: A, most important; B, middle; C, least important.
SOURCE: NRC Working Group.

nology for health care, biotechnology is part of a toolbox of enabling technologies for drug discovery and development. For large Japanese companies (LJCs) using biotechnology for health care, biotechnology is critically important to generate new products and in some cases to diversify into completely new business areas.

CEOs of LUCs and LJCs understand that a global orientation is essential to ensure long-term competitiveness and financial returns. The United States is the world's largest pharmaceuticals market; as such, LJCs need to penetrate it.[40] Another factor that is just as important in stimulating linkages between LJCs and SBFs is a desire to access products and technology developed in the United States. From the perspective of an SBF, linkages are essential under current conditions in order to obtain capital needed to support R&D-intensive operations and gain access to the Japanese market. Meanwhile, LUCs are developing strategies that are more narrowly focused from a technical perspective. LUCs, seen by some as less successful than SBFs in developing their own technology and less adept than LJCs in obtaining

---

[40] No LJC has dominant market share in Japan; LJCs are aiming to expand in global markets.

technology developed by others, may seek to develop their own Japanese subsidiaries in order to strengthen their distribution and marketing capabilities in Japan and, over the long term, to improve access to Japanese technology.

In considering the driving forces behind linkages between LJCs and SBFs, capital requirements deserve special attention. For SBFs, infusion of capital is a matter of survival. LUCs, driven by pressures to produce earnings per share, tend to focus on defensive strategies_mergers and linkages for diversification into new markets and support of existing areas. LJCs, in contrast, have the capital resources needed by SBFs and the drive and long-term focus to establish linkages to complement plans for expansion through new technology and products.[41]

Technology linkages provide SBFs with a means of survival through capital infusions. For LJCs, technology linkages with SBFs are an important component of a global growth strategy. LUCs, in contrast, take a more defensive approach to technology linkages, particularly with Japanese firms. As U.S. SBFs turn to Japanese capital, the LJCs gain access to U.S. technologies. The LJCs are well equipped to exploit the products and technologies developed in the United States. One possible approach to address this issue would be for the U.S. government to establish incentives to reward SBFs that bring new technology to the commercialization stage, or work with other U.S. companies to do so.

In addition, differences in patent systems may encourage these trends.[42] The first-to-invent system used in the United States gives stronger protection to the inventor than does the first-to-file system used in Japan and most other countries. While the United States maintains strong intellectual property protection for biotechnology, obtaining patent protection is a time-consuming and uncertain process. U.S. researchers in SBFs, universities, and research institutes may be ready to discuss their work at an early stage, before patent protection is granted, with individuals from LJCs who shop the world for new technologies and who have the option of applying for patent rights in Japan based on knowledge of work done elsewhere.[43] SBFs have limited resources for R&D; they cannot pursue all new developments in research, and the "overflow" presents opportunities for LJCs and LUCs.

---

[41] Due to differences in accounting rules, it costs more for a U.S. company to purchase another U.S. company than it does for a foreign-based firm to do so.

[42] See Committee on Japan, National Research Council, *Intellectual Property Rights and U.S.-Japan Competition in Biotechnology: Report of a Workshop* (Washington, D.C.: National Research Council, 1991).

[43] Joshua Lerner argues that the *breadth* of patent protection is another issue for small U.S. biotechnology firms. His analysis shows that small U.S. firms that were awarded broad patent rights in their early years are more likely to succeed. See Joshua Lerner, "The Impact of Patent Scope: An Empirical Examination of New Biotechnology Firms," Center for International Affairs, John F. Kennedy School of Government, July 1991.

While LUCs with operations in Japan have a thorough understanding of the regulatory and health care environment there, SBFs often lack understanding of these factors. Another factor influencing the nature of relationships between SBFs and LJCs is the fact that SBFs must disclose their technology, publicly or to their potential partners, in order to obtain capital investments.

The internal focus of LUCs on U.S. technology, the pressure to perform in the short term in order to achieve high stock valuations, the large domestic market, and the U.S. regulatory process have made it difficult for them to take advantage of opportunities for accessing technology developed in other parts of the world. Although some U.S. pharmaceutical companies have established R&D facilities in Japan, much of the work carried out there is development rather than research.[44] The NRC working group is aware of only a small number of U.S. pharmaceutical licensing executives stationed in Japan to look for licensing opportunities from Japanese companies.

LJCs, in contrast, invest considerable resources in following the world scientific literature, specialized conferences, and research around the world. Due to the comparative weakness of Japanese basic research in universities and the limited numbers of Ph.D.'s produced, Japanese corporate scientists rely heavily on the work of scientists from the United States to elucidate fundamental theories as a basis for more applied research.

SPECIAL CHARACTERISTICS OF JAPANESE INVESTMENT IN THE U.S. BIOTECHNOLOGY INDUSTRY

As noted above, large Japanese pharmaceutical companies along with chemical and food companies have been the major investors from the Japanese side in the U.S. biotechnology industry. In addition, a greater proportion of nonpharmaceutical Japanese, as compared to U.S., companies are attempting to diversify into health care. The traditional impetus for international market expansion by the Japanese pharmaceutical companies has been augmented by conditions specific to their domestic market_fierce competition and government-imposed drug pricing. Acquisition of large U.S. pharmaceutical firms is beyond the capabilities of most of even the largest Japanese pharmaceutical companies. Their general unfamiliarity with the U.S. regulatory process and their lack of managers experienced with international operations also are limiting factors to their penetration of the U.S. market.

---

[44]See *Survey of Direct U.S. Private Capital Investment in Research and Development Facilities in Japan* (Washington, D.C.: National Science Foundation, 1991). The survey uncovered 71 U.S. firms with R&D facilities in Japan, a few in the pharmaceutical industry. Upjohn expects to have a research staff of 400 working at its new laboratory.

As a result, Japanese pharmaceutical companies are following long-term incremental strategies in their investments in the U.S. biotechnology industry. Threefold objectives_namely, (1) acquisition of limited product rights to strengthen the company's position in the home market, (2) acquisition of technology to strengthen the company's technology base, and (3) initiation of a relationship with a U.S. biotechnology firm as part of a long-term education process_also represent steps in a sequence. [45]

The acquisition of Gen-Probe, Inc. by Chugai Pharmaceutical Company illustrates this process. The relationship formally began in April 1988 through a $15.5 million transaction involving Asian distribution rights and technology and product development. Approximately 18 months later, Chugai took another step by acquiring Gen-Probe. This provided Chugai with a platform to expand its U.S. presence, consistent with the establishment of several licensing agreements and a joint marketing venture with Upjohn.

Fujisawa Pharmaceutical Company has followed a similar strategy. After an initial investment for a 30 percent stake in Lyphomed in 1984, Fujisawa completed the acquisition in late 1989. Although Lyphomed is a generic drug company rather than a biotechnology company, its dominant presence in the hospital-based market for injectables and its access to certain novel products make it a unique and attractive company. Fujisawa has also forged several technology linkages and product relationships in biotechnology and in late 1987 bought out its joint effort with SmithKline, Fujisawa SmithKline.

Yamanouchi Pharmaceutical Ltd. has made an unusual entry into the U.S. market. Following an apparent takeover attempt of Shaklee Corporation by Irwin Jacobs, Yamanouchi moved expeditiously in March 1989 to acquire the company for $395 million, presumably to protect the 78 percent interest in Shaklee Japan KK, which it had acquired the previous month. Although the Shaklee acquisition is outside the pharmaceuticals sector, it demonstrates that Japanese pharmaceutical companies can act quickly when necessary to shore up their strategies. Yamanouchi's strategy appears to be broader; the company is planning a new research center in the United States and also has joint development efforts with several biotechnology companies, including Genetics Institute and Alteon.

Kirin represents an example of investment motivated by a desire for diversification into new product areas. Kirin's long-term vision of 1980 contemplated diversification into drugs. It reckoned that biotechnology

---

[45] See Booz-Allen & Hamilton, "Foreign Investment in Bay Area Bioscience," 1991. This report provides data to document the conclusion that foreign investment accounted for 59 percent of the $6.3 billion invested in the Bay Area in biotechnology between 1975 and 1990. Excluding the $2.1 billion acquisition of a 60 percent stake in Genentech by Roche, foreign capital accounted for 38 percent of the total inflow. The report shows that the vast majority of foreign investment came from Europe, while Japanese sources provided 15 percent.

provided a special entree, due to the conservatism of the large Japanese pharmaceutical firms. Over the past decade, Kirin has nurtured relationships with Amgen to develop EPO and has established the La Jolla Institute for Allergy and Immunology. Kirin can afford to look far into the future, plotting a strategy to achieve technological interdependence.[46]

In all cases, acquisitions require the building of personal relationships. These are usually solidified in the early stages of a business relationship that may involve simple technology licensing from a U.S. firm.

Large Japanese firms see penetration of European markets as a basis for further expansion. Such investments could also provide a foundation for the future acquisition of U.S. firms. A good example of Japanese expansion in Europe is Yamanouchi. In September 1990, Yamanouchi opened a cell biology research unit outside Oxford in the United Kingdom. This is Yamanouchi's first international research center and was designed to serve as a bridge between academia and industry. Yamanouchi already operates a manufacturing facility in Ireland and has a presence in most European countries. In addition, it completed the takeover of Gist-Brocades' pharmaceuticals division in February 1991, further strengthening its European presence.

Takeda, the leading Japanese pharmaceuticals company, has a long history in Europe. Takeda has established a series of joint ventures with many European companies. In 1978 it formed a French joint venture with Roussel UCLAF, another in 1981 with Grunenthal, and in 1982 yet another joint venture with Cyanamid Italia. In 1988 Takeda established an R&D center in Frankfurt. There were reports in mid-1991 that it was one of several Japanese companies interested in equity investment in Gedeon Richter, a premier Hungarian pharmaceuticals company.

The privatization effort under way in Hungary has attracted a lot of attention and illustrates Japanese interest in acquiring a stake in Eastern Europe. Nomura Securities is advising the State Property Agency of Hungary in the privatization of Richter, assisting in the search for an equity investor. As of June 1991, seven Japanese companies, including Takeda, Fujisawa, and Sankyo, were reportedly exploring investments in Richter, in addition to European and U.S. companies. Another of Hungary's premier pharmaceutical companies, Egis Pharmaceuticals, signed a funding and cooperation agreement with Japan Tobacco in May 1991. This comes on the heels of several investments by Japan Tobacco in U.S. biotechnology companies (e.g., Mycogen, Isis Pharmaceuticals, Cell Genesys) and is part of a global expansion and diversification strategy in pharmaceuticals and agribusiness.

---

[46]Noboru Miyadai, "Kirin Biru_Beikoku Bencha Kigyo to no Goben ni Yoru Iyakuhin Kenkyu Kaihatsu no Jissai" (Kirin Beer_Realizing Pharmaceutical R&D Through a Joint Venture with a U.S. Start-up), *Business Research* (in Japanese), March 1989, p. 25.

Aside from strategic investments, the interest of financially motivated Japanese institutional investors in U.S. biotechnology stocks has been limited. There are good reasons why this is the case. Japanese institutional investors as a group are risk averse and interested primarily in financial instruments with yield, such as bonds and dividend-paying stocks. This investment philosophy is consistent with the fact that the performance of portfolio managers is typically based on their ability to generate current income while preserving the principal, rather than taking risks that may result in capital appreciation. This investment philosophy bias toward current income explains the reported practice of Japanese portfolio managers to buy stocks immediately prior to dividend distribution and sell them shortly thereafter.

There are other reasons that argue against widespread ownership of U.S. biotechnology stocks by Japanese institutional investors. Investing in U.S. equities is a relatively recent phenomenon for Japanese institutions, which have traditionally invested in U.S. government obligations and investment-grade corporate bonds. Most Japanese portfolio managers responsible for investing in U.S. equities are relatively junior and regard that position as a stepping stone to more significant fund management assignments or other positions in the organization. Such short-term assignments do not permit the acquisition of necessary experience in the unique world of biotechnology stocks. In addition, the relative lack of liquidity of the stocks of most small biotechnology companies makes them unsuitable for the large funds typically managed by Japanese institutions.

Venture capital is increasingly talked about in Japan, but a significant domestic venture capital industry has yet to emerge. For several years some Japanese corporations have made investments in U.S.-based venture capital funds, aiming at both significant returns and a window on technology.[47] Japan Tobacco and Snow Brands are such examples of Japanese companies involved in U.S. venture firms. By contrast, there has been little interest by Japanese institutional investors in this area. It seems likely that there will be efforts to establish Japanese-managed or partnered venture funds for the purpose of investing in U.S. biotechnology companies. JAFCO, a division of Nomura Securities, has been the most notable participant in this area.

---

[47] According to Venture Economics, publicly reported minority equity investments by Japanese corporations in U.S. venture businesses increased 45 percent in 1989 over 1988 to a level of $320 million, and levelled to $306 million in 1990. (Communication with Venture Economics, 1992.) By far the major focus of this investment was in the electronics field. According to *Corporate Venturing News*, vol. 4, no. 7, May 4, 1990, investments in medical and health care made up only about 8 percent of the total number and 2 percent of the dollar value of these investments in 1989.

TECHNOLOGY LINKAGES BETWEEN JAPANESE COMPANIES AND U.S. UNIVERSITIES AND NONPROFIT RESEARCH INSTITUTIONS

While company-to-company linkages provide the most direct mechanism for transferring technology for the purposes of commercialization, there is good reason to examine the growing linkages between Japanese firms and U.S. universities and nonprofit research institutions. The NRC biotechnology working group concluded that Japanese funding of research at U.S. universities and contract research with U.S. university professors together are the second most important vehicle (in terms of potential future impact on the industry) for technology transfer. These linkages take a variety of forms_including funding for departments or chairs, participation in corporate liaison programs, licensing from university technology transfer offices, contract research by individual professors, and training of Japanese researchers in scientific disciplines associated with biotechnology at major U.S. research institutions.

There is no comprehensive data base to measure these interactions, and in many cases there is only anecdoctal evidence of an apparent trend toward rapidly expanding linkages.[48] In the mid-1980s the General Accounting Office found that U.S. universities received $16.4 million from foreign sources, of which $2.6 million was contributed to biology departments and $8.4 million to departments of medicine.[49] There is evidence of a significant increase in such funding in the recent past, illustrated by Shiseido's $85 million investment in Massachusetts General Hospital. Appendix B provides a summary of major technology linkages between Japanese companies and U.S. universities. These investments likely reached the level of $50 million in 1989 and are increasing at a rate of 25 percent per year.

One Japanese expert estimated that 10 years from now, Japanese companies will be spending $3 billion annually in U.S. universities.[50] In 1990 the Japan Productivity Center sponsored a survey of how Japanese firms are using foreign universities and research centers. The most important relationship,

---

[48]The General Accounting Office (GAO) carried out a study of foreign funding of U.S. university research and concluded that it constituted only a small part of all university R&D expenditures ($74 million or about 1 percent of total university R&D expenditures). The study also found that foreign funding was concentrated in a few of the nation's largest research universities. Japan sponsored more R&D than any other country, about $9.5 million in 1986. See GAO, *R&D Funding: Foreign Sponsorship of U.S. University Research*, Washington, D.C., March 1988.

[49]GAO, ibid., p. 16. Eight universities received $4.7 million in foreign funding for research in areas such as clinical testing of pharmaceuticals, neuropsychiatric research, and radiology.

[50]See comments by Konomu Matsui, reported in "Japanese Forging Ties with U.S. Universities," *Research-Technology Management*, January 1991, p. 3.

according to the respondents, was the use of a university as a subcontractor for basic and pioneering R&D (39 percent), followed by joint partnership with a university for such projects (30 percent). The survey confirmed that Japanese companies, like their U.S. counterparts, are seeking to establish linkages with universities that will benefit their own businesses.[51]

The list of technology linkages between Japanese companies and U.S. universities in Appendix B confirms that the focus has been on new technology development. A prominent example, mentioned above, is the investment of $85 million by Shiseido to establish the world's first comprehensive cutaneous biology center at Massachusetts General Hospital, Harvard University's largest teaching hospital. In this case a sponsored research agreement was negotiated that outlines patent protection, salaries, direct and indirect costs, and relationships with other sources of funding. According to individuals at Massachusetts General Hospital, a sponsored research agreement provides more insurance for the U.S. organization than would be the case if a "gift" were made by a Japanese company to an individual professor. Gifts, often not covered by university policies, are sometimes made directly to an individual professor who is part of a publicly funded research program. Whether the funding comes from a U.S. or Japanese company, the company is always interested in gaining intellectual property rights. Often this is accomplished by filing a patent application prior to publication of research results, so that the corporate sponsor's intellectual property rights are protected.

Technology linkages between Japanese companies and U.S. research institutions must be seen in a larger context_that of the relative comparative advantages of the two countries. In the past few years concerns have been raised about declining award rates by the NIH and NSF to U.S. researchers. Both agencies have concluded that there is much potentially valuable science represented in the applications that have gone unfunded. In addition to advocacy for continued support for basic research, a number of U.S. science policy leaders have begun to call for an increase in nondefense R&D, with an eye toward strengthening the competitiveness of U.S. industry. Steps have also been taken to improve technology transfer from the national laboratories to industry, and an effort has been made to provide a preference for transfers of technology to U.S. firms.

In a period when research funds in the United States are constrained and priorities are under discussion, the number of foreign researchers in U.S. university and nonprofit research laboratories is growing. In 1988, according to statistics prepared by Japan's Ministry of Justice, 52,224 Japanese researchers went to the United States, while 4,468 U.S. researchers

---

[51] Ibid.

visited Japan.[52] A number of studies have explored the growing importance of foreign-born scientists and engineers to R&D in the United States. The excellent and open laboratories in the United States attract researchers from around the world.[53]

This is certainly true with respect to the exchange of Japanese and U.S. researchers in biotechnology. While only comparatively small numbers of U.S. researchers are going to work in Japanese laboratories, about 30 percent of the U.S. individual researchers who spent at least two months in Japanese government-supported programs in Japan identified themselves as working in the field of life sciences. Almost two-thirds of the Japanese researchers who spent more than 1 month at U.S. national research institutes in 1988 were reportedly carrying out research in biotechnology.[54] It is estimated that there are 450 researchers from Japan at NIH, out of a total of 1,800 foreign researchers.[55] While data are inadequate to provide an accurate estimate of the exchange of U.S. and Japanese researchers in biotechnology, it appears that biotechnology is a significant area of mutual interest.[56]

There is no easy way to calculate the gains or losses to the United States. Close interaction with a senior scientist represents access to years of funding and a network of researchers. Foreign researchers contribute to the work of the laboratories they visit. But the full costs of training are not covered by stipends or salary support. Japanese researchers, particularly those from private companies, usually return to their home country laboratories. In a few cases, however, talented young Japanese scientists have said that they would be unable to pursue creative research in Japan.[57]

Steps can be taken to expand the number of U.S.-born students, including women and minorities, who pursue careers in science. Programs of Japanese-language training for technical personnel and expanded fellowship opportunities may, over time, increase the number of U.S. researchers who

---

[52] More than 113,000 Japanese researchers went to countries around the world, while Japan received 68,000 (59,000 from developing countries).

[53] While U.S. university research remains comparatively strong, researchers are voicing complaints about inadequate funding, administrative constraints on multidisciplinary research, and other problems.

[54] The total number of U.S. researchers covered in the survey conducted by NSF was 94.

[55] Data provided by NIH. About two-thirds of all foreign researchers at NIH receive some support from NIH; it estimated that roughly the same percentage of Japanese researchers receive support. It should be noted that this support covers salaries, but not the true and full costs (or contributions) of using the facilities and interacting with permanent staff.

[56] The 1988-1989 Annual Report of the Japan Society for the Promotion of Science indicates that only two of the 75 American postdoctoral researchers in Japan were working on topics related to biology. Of the 19 Japanese postdoctoral students in the United States in the same period, eight were engaged in biomedical research. See pp. 70-74.

[57] Nobel prize winner Susumu Tonegawa is one example.

study and work in Japanese laboratories. Until the quality of Japanese basic research in the life sciences improves, however, incentives for U.S. scientists to work in Japanese universities and national research laboratories will remain limited. For that reason, meaningful access to Japanese biotechnology R&D must include opportunities to interact with corporate laboratories and industry-led R&D consortia.[58]

The 1988 umbrella agreement between Japan and the United States for cooperation in science and technology states that increased cooperation is a goal. Cooperation in life sciences, including biotechnology, has been identified as a priority area. Experience over the past few years would lead one to question whether the umbrella agreement is a potent instrument for fostering research exchange in biotechnology.[59] A team of U.S. experts traveled to Japan in 1991 to assess the status of Japanese bioprocessing; their visits focused primarily on R&D facilities.[60] The major impetus for expanding collaborative R&D efforts has come rather from individual agencies, such as the agreement between NSF and the Ministry of Education to promote bilateral seminars in biotechnology and other fields.[61] The Science and Technology Agency of Japan began a cooperative research project with the U.S. National Science Foundation in biotechnology.[62]

For universities and national laboratories supported with public funds, important questions have been raised concerning reciprocity. Recent public debates have focused on industrial liaison programs that include large numbers of foreign companies, research sponsored by foreign companies at U.S. universities, and the growing number of Japanese researchers in the nation's premier public sector biotechnology laboratories. These debates have drawn attention to the question of whether the end result will be to build a formidable competitor in Japan's biotechnology industry. It is not surprising that

---

[58] Committee on Japan, National Research Council, *Expanding Access to Precompetitive Research in the United States and Japan: Biotechnology and Optoelectronics* (Washington, D.C.: National Academy Press, 1990).

[59] The most significant exchanges are worked out independently by such agencies as NIH or take place through individual exchanges, rather than being fostered under the umbrella agreement.

[60] Japan Technology Evaluation Center (JTEC), "Bioprocess Engineering in Japan," forthcoming.

[61] Two such seminars have been sponsored in recent years, each involving about 10 researchers from each country. In addition, a significant number of NSF's cooperative science programs with Japan have focused on biotechnology-related topics. There were a total of 56 seminars in all fields in the 1989-1990 period. Only a small number of U.S. researchers on long-term stays in Japan, however, are working in biotechnology-related areas, according to NSF.

[62] See "Baio de Bei to Kyodo Kenkyu" (Joint Research with the United States in Biotechnology), *Nihon Keizai Shimbun* (in Japanese), January 5, 1991, p. 13. The collaboration involves Michigan State University and will be focused on environmental applications of biotechnology.

Japanese companies are building strategies to access basic research in U.S. universities, in view of the significant costs that would have to be incurred to establish comparable programs in-house and the relative weakness of Japan's own university labs.

In an effort to learn more about the linkages between U.S. universities and Japanese corporations, the NRC working group conducted a pilot survey in the spring of 1991. The survey was sent to 23 of the largest university biotechnology centers. Responses from 18 of the centers indicated that the value of research contracts with Japanese companies is small and that little technology is licensed to Japanese companies but that the number of Japanese visitors and researchers is significant. Confirming previously mentioned trends, there is no case in which the number of researchers from the U.S. university biotechnology centers going to Japan approaches the number of visitors from Japan. It should be noted, however, that most respondents indicated that they did not have complete information about linkages developing across their university.

The issue of Japanese participation in research at U.S. universities is complex. U.S. university officials say that participation by Japanese companies often comes after U.S. companies have declined to get involved. In view of the federal budget crisis and the exponential growth in R&D in the life sciences, it seems likely that Japanese involvement will increase in the years ahead. While some believe that restrictions are needed to protect U.S. competitiveness, a more viable approach may be for universities themselves to build more coherent strategies. In view of the growing public concern, it may be appropriate for U.S. institutions to develop guidelines that permit the continuation of foreign participation while ensuring academic freedom and timely dissemination of research results.

Questions of reciprocity also arise in the context of participation in international conferences and dissemination of research results through professional journals and data bases. There is no satisfactory way to judge the numbers of Japanese researchers attending conferences in the United States or other locations or to draw firm conclusions about their contributions to professional organizations as paper presenters and program organizers (as contrasted to registrants who come to listen). Some large organizations, such as the Federation of American Societies for Experimental Biology (FASEB), whose annual meetings draw 15,000 to 20,000, would find it difficult to make statements about participation by individuals from Japan. FASEB collects data on the number of "registrants from abroad" (only about 50 from Japan in 1990), but there is no way to know how many of the Japanese researchers currently working in the United States might register using their current institutional affiliation.[63] In the cases of the Association

---

[63] Data provided by FASEB.

of Biotechnology Companies (ABC), there were eight Japanese attendees at the 1989 annual meeting (out of a total of 475) and five Japanese attendees at the 1990 meeting (which was attended by 500 individuals).[64] Some U.S. professional organizations have few conferences or events that involve foreign participation.[65]

On the other hand, there is good reason to believe that professional conferences and meetings with a more specialized focus in some subfields of biotechnology, particularly those held in the Pacific region by international unions, attract good participation from Japan. For example, the major interaction of the American Chemical Society (ACS) with Japan is not in its annual meeting but in the International Chemical Congress of Pacific Basin Societies, which the society cosponsors. Of the 6,000 registrants at the 1989 congress in Honolulu, Hawaii, more than 3,400 were Japanese chemists.[66] Likewise, Japanese attendance at smaller biotechnology-related meetings organized by FASEB may well be much stronger than at the association's annual meeting. One can find evidence to support this perspective by observing the strong participation of Japanese chemical engineers in conferences that feature new research on safety-related areas.[67]

The best example of a U.S. professional organization involved in biotechnology where there is significant foreign participation is the American Society for Microbiology (ASM), which also collects the most relevant data.[68] More than 25 percent of ASM's members come from abroad, and they are seen as making substantial contributions to the organization. Statistics from the ASM are especially relevant, since members must have at least a bachelor's degree in microbiology. Of its almost 35,000 full members, about 1,000 are Japanese. The areas of expertise most frequently cited by the Japanese members are molecular biology and fermentation. Nonmember subscriptions to journals are unusually high for the Asian region, and Japanese members' comparatively high subscription rates to the journal *Clinical Microbiology* reflect their strong interest in pharmaceutical-related applications. Foreign authors are significant contributors to ASM's more than 10 journals. ASM is a good example of an American professional association that is consciously charting an international course. Japanese scientists, from industry as well as academe, participate significantly not only in attending meetings but also in conference planning, authoring papers, and subscribing to publications (see Tables 7a, 7b, and 8).

To better understand the internationalization of biotechnology, it would

---

[64] Data provided by ABC.
[65] One example may be the American Institute for Chemical Engineering.
[66] Data provided by ACS.
[67] Comments from the staff of the Biochemistry Union.
[68] ASM generously cooperated with the NRC working group in providing the detailed statistics cited here.

TABLE 7a ASM Full Membership by Region

	1985	1986	1987	1988	1989	1990
U.S. full members	22,842 81.01%	23,510 80.47%	24,103 79.82%	25,100 78.95%	26,203 78.09%	26,725 77.29%
Asia full members	1,288 4.57%	1,382 4.73%	1,458 4.83%	1,573 4.95%	1,748 5.21%	1,906 5.51%
Total full members	28,184	29,216	30,197	31,792	33,556	34,578

˘ Japanese full members constitute approximately 50% of the Asia region. (Currently, there are 1,000 Japanese full members out of a total of 35,000.)
˘ Full members hold at least a B.S. in a microbiology or a related science.

SOURCE: ASM, Washington, D.C.

TABLE 7b 1990 ASM Full Members by Self-Identified Divisions

	No. in U.S.	No. in Asia	Total No. in Division from Entire Membership
Molecular biology	3,218 79.54%	243 6.01%	4,046 11.70%
Fermentation	810 76.13%	83 7.80%	1,064 3.08%
Clinical microbiology	4,626 78.83%	241 4.11%	5,868 16.97%
Medical microbiology	1,032 74.46%	88 6.35%	1,386 4.01%

˘ Percentages are based on total number of members in each division.
˘ Divisions here are those in which the greatest number of Japanese members identified themselves.

SOURCE: ASM, Washington, D.C.

TABLE 8 ASM Journal Subscriptions and Publication Acceptances for 1990

	U.S. Full-Member Subscription	Asia Full-Member Subscription	U.S. Nonmember Subscription	Asia Nonmember Subscription	U.S. % Accepted in Journal	Asia % Accepted in Journal
Medical Microbiology	3,947	342	945	351	56.43	7.68
Applied & Environ. Microbiology	4,331	388	1,056	710	55.34	9.32
Molecular & Cell Biology	3,724	289	818	268	78.41	4.15
Clinical Microbiology (review)	5,479	452	249	90	80.95	0.00
Infection and Immunity	3,549	374	771	414	61.81	9.05
Systematic Bacteriology	494	213	311	178	39.54	19.77
Journal of Bacteriology	3,390	489	1,354	751	62.18	8.61
Clinical Microbiology (diagnostic)	8,354	621	1,138	405	55.54	8.99
Journal of Virology	2,677	248	808	374	67.71	5.28
General Microbiology (review)	6,778	618	1,313	503	66.67	8.33

" Number of Japanese authors contributing from the Asia category is high.
" Percentage of foreign contributions to ASM journals during 1986-1990 was steady (38%) but increased in *Applied and Environmental Microbiology*, *Systematic Bacteriology*, and *Journal of Clinical Microbiology*.
" Most subscriptions in the Asia category are from Japan. Interest of Japanese scientists in identifying new microorganisms is very high. This is reflected in the high subscription rate for *Systematic Bacteriology*, which focuses on nomenclature.
" ASM is increasing the number of foreign scientists on its editorial and advisory boards.

SOURCE: ASM, Washington, D.C.

be useful to gather more comprehensive statistics on participation by foreign scientists and engineers in conferences, as organizers and panelists as well as attendees, and in specialized journals. Comparisons to conferences held in Japan and to Japanese-language publications would provide a basis for drawing conclusions about reciprocity. Unless universities and professional associations carry out such studies and cooperatively analyze data gathered by different institutions, the policy debate will be influenced by anecdoctal evidence and inadequate statistics.

EXAMPLES OF TECHNOLOGY LINKAGES_MULTIPLE PURPOSES AND MECHANISMS

To better understand the multiple mechanisms that typically make up U.S.-Japan linkages in biotechnology, it is important to look beyond aggregate data to specific examples. Four examples of U.S.-Japan technology linkages have been examined for this report, and full case studies are included in Appendix A. The cases are Kirin's operating joint venture with Calgene in seed potatoes, Monotech's licensing and marketing relationship with Showa-Toyo Diagnostics (STD) in cancer diagnostics,[69] Kirin and Amgen's joint venture to develop and market EPO and granulocyte colony-stimulating factor (G-CSF), and the lease-swap agreement that allowed Hitachi Chemical's U.S. subsidiary to build an R&D lab on the campus of the University of California at Irvine in return for use of space in the building by the University's Department of Biological Chemistry. These examples illustrate much of the range of technologies, linkage mechanisms, partners, and markets that currently make up U.S.-Japan biotechnology linkages.

The studies reinforce the inference drawn from aggregate statistics and anecdotal evidence that the objectives for Japanese partners are largely technological and that the U.S. partners are typically motivated by financial considerations.

In addition to illustrating the motivations of the partners, the cases also detail the process of forming and managing biotechnology linkages. Each case is unique, but some common themes come into focus, including the gradual process of building relationships over time and the use of multiple channels to establish linkages.

Some emerging questions are also discernable. For example, under what circumstances do linkages_even those that have benefited both sides_ have a continuing rationale beyond the success of the first products in the market? U.S. biotechnology firms experiencing success may have greater bargaining power than in the past relative to larger corporate partners_

---

[69] Not the real names of the companies.

Japanese and otherwise. The question is whether the structure of linkages can be modified to reflect changes in the circumstances and interests of the partners. Another important question concerns the long-term significance for U.S. academic research and for U.S. competitiveness of new efforts on the part of large Japanese companies to establish closer relationships with the U.S. biomedical research community in universities and other academic settings.

What conclusions can be drawn about the impacts on the partners, both short and long term? Of the four cases presented here, two were launched fairly recently, which makes it difficult to assess even short-term impacts. In the other two cases the positive short-term impacts on both sides have been obvious and substantial.

The short-term benefits to Japanese partners have centered on products. Japanese companies have been able to gain a foothold in commercial biotechnology by licensing products and technology developed in the United States. It is important to remember that, because of the time required for approval of pharmaceutical products, "short-term" benefits may take a number of years to materialize.

In several of the cases the Japanese partner also focuses on expected longer-term benefits, such as diversification or a migration to more innovation-intensive strategies in existing businesses, in some instances by building research capability in biotechnology. In some cases the stated aim of many of the Japanese companies locating R&D facilities in the United States is to eventually conduct independent advanced biotechnology R&D in the United States. In a general sense, Japanese partners are motivated by the desire to establish a stronger global presence, although the cases studied here are unlikely to result directly in increased sales of products developed in Japan. As the two cases involving Kirin Brewery show, the pursuit of shorter- and longer-term benefits is by no means mutually exclusive, and perhaps linkages that bring both sorts of benefits are ideal from the standpoint of Japanese companies. But the greater product focus of Monotech-STD and the greater capability-building emphasis of the Hitachi Chemical/ UC Irvine relationship may illustrate general differences between alliances forged with U.S. firms and with American research institutions.

Clearly, Japanese companies have been able to use linkages with U.S. institutions to build technological and marketing capabilities in commercial biotechnology_regardless of whether particular linkages are maintained or dissolved. However, the eventual payoff of these capabilities and the extent of long-term benefits to the Japanese partners remain to be seen.

From a U.S. perspective, a common thread in the calculations of companies_particularly small biotechnology firms_as well as universities is the need for capital to support a world-class R&D effort. In an environment of constrained federal R&D budgets and impatient investors in large U.S. corporations, the prospect of investment from Japan and Europe presents an opportunity that U.S. institutions will consider seriously. The case studies show that under certain circumstances U.S. partners have been able to utilize linkages with Japan as part of an overall strategy for growth. Linkages have contributed critical financial resources to some U.S. firms. It appears that the quality of the American partner's technology is crucial in determining the potential of linkages to bring substantial benefits in the short term and that superior technology must be combined with clear strategic vision on both sides to realize its commercial potential. U.S. partners have also realized technological benefits from linkages with Japanese companies, though these are seldom consciously pursued.

Market access also is an issue for American partners. For the stronger biotechnology firms, the development of linkages with Japanese companies has opened opportunities in the Japanese market for expanded sales, mostly licensed sales.

Over the longer term, and parallel to the expectations of Japanese firms setting up R&D facilities in the United States, there is also a prospect that the linkages will serve to improve access to biotechnology developed in Japan. A handful of U.S. biotechnology companies are now monitoring Japanese technology, funding research, and conducting clinical trials in Japan. It is important to note that Japanese companies are strongly oriented toward technology development rather than fundamental science. This asymmetry in the biotechnology R&D systems of the two countries will make it necessary for the U.S. partners to consciously develop strategies to access the applied technology, particularly production technology, developed in Japan.

Building the capability to enhance the long-term benefits of linkages with Japan has been vigorously pursued by some U.S. partners but doing so is often difficult. This is understandable given the financial and human resource constraints that have characterized most U.S. biotechnology companies. Yet devoting resources and attention to leveraging linkages with Japan to obtain technology and a foothold in the Japanese market may be an important focus for long-term growth and survival.

Readers should keep in mind that detailed studies often touch on sensitive issues that parties connected with linkages are reluctant to discuss, particularly when real names are used. It is difficult to elicit information about cases that are clearly unsuccessful, or where there are hard feelings, under any circumstances. Further, although the cases contain examples of U.S.-Japan linkages in biotechnology aimed at human therapeutics, diagnostics, and agricultural biotechnology as well as a range of institutions as partners, none involve one of the major Japanese or American pharmaceutical companies. There have been few acquisitions of U.S. biotechnology firms by Japanese companies, and it was not possible for the NRC working group to prepare a study on that particular technology transfer mechanism. Thus, these cases are not ideally representative, nor do they comprehensively illustrate the factors that can lead to asymmetrical benefits or outright failure. Still, care has been taken to explore cases in which a range of business and technical issues arise.

4

Prospects for the Future

What are the prospects for the decade ahead? What are the general trends, and what are the factors that could introduce significant changes in these trend lines? The current picture is one of U.S. predominance in biotechnology and continuing technology transfer to Japan, based on the unique role of innovative U.S. biotechnology firms and U.S. universities as the dynamic foundation for commercial R&D.

The overall lead that U.S. firms enjoy today in biotechnology R&D is, however, insufficient to guarantee future competitive success. Japanese firms present a significant competitive challenge, one likely to grow in the years ahead. Japanese companies involved in biotechnology believe that it is critical technology, and they are developing broad-based strategies, in cooperation with their government, to use biotechnology as a base to move into diverse new business areas.

The NRC working group expects biotechnology to surge in significance in the decade ahead. The number of U.S. companies will continue to expand overall, despite some consolidation. The structure of the industry will change, showing less concentration in the health care field.

The U.S. today has some important advantages over Japan_the involvement of university researchers in new and innovative companies and a vibrant venture capital market. In 1991 nearly $4 billion in equity capital was raised in the United States. This new financing provides U.S. biotechnology firms with a stronger hand in negotiating linkages. We are witnessing the creation of a new paradigm_small U.S. biotechnology companies are linking up with larger U.S. companies and U.S. universities. This triad arrangement holds the promise of strengthening the foundation for the industry through faster commercialization.

Estimates of the biotechnology market in the United States, Japan, and Europe in the year 2000 vary considerably. According to an analysis by the North Carolina Biotechnology Center, the Japanese market will see sales of $30 billion in the year 2000, exceeding sales in the United States or Europe. Other estimates by CEOs project sales in the United States of $50 billion by the end of the decade.[70] Other indications of the importance of the industry are that (1) sales of biotechnology products by U.S. industry are nearing $3 billion, (2) the equity value of public companies in the biotechnology sector grew by nearly 40 percent in 1990, (3) the biotechnology sector was the number-one performer for 1990 according to Dow Jones.[71] (See Figure 3.)

Looking to the future, U.S. biotechnology firms will continue to proliferate. While consolidation among companies (and acquisition of some) will continue, the rate of start-ups in the United States will exceed the rate of consolidation; thus, net industry growth will continue. However, industry growth (by numbers of companies) may slow. For example, during the 1980s, there were 50 to 75 start-ups each year; during the first half of the 1990s the rate is likely to decrease.

Small biotechnology firms will probably form strategic alliances with large foreign and domestic companies at earlier and earlier stages in their growth/existence. During the 1980s, many biotechnology firms formed strategic alliances with corporate partners in periods when capital was expensive and becoming increasingly difficult to obtain. This trend caused most partnering to be done by the larger biotechnology firms whose principal purpose was to obtain capital. As we move through the 1990s, the general purposes of these alliances will change, as will their character and purpose:

˘ Alliances will be formed earlier and earlier in a biotechnology firm's existence. Today we see biotechnology firms forming alliances shortly after formation, even sometimes as a precursor to obtaining venture capital. These alliances are a de facto validation of the new biotechnology firms (thereby encouraging venture and other investment) and also bring specific technical resources to both sized companies. For the bigger players, the technology is increasingly unavailable to them in other forms. (A Nobel-prize-caliber scientist with cutting-edge technology for further development may be willing to create his or her own company and partner the technology for further development with a large corporate partner but may be unwilling to become an employee of a large company.) Thus, a new paradigm for

---

[70] See Burrill and Lee, *Biotech 91*, op. cit., p. 30.
[71] See the newsletter *Washington FAX*, January 14, 1991.

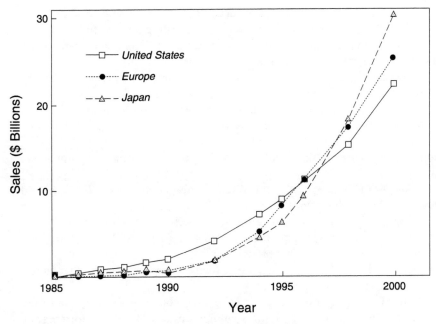

FIGURE 3 Biotechnology sales, 1985-2000. SOURCE: Mark Dibner.

innovation is occurring: cutting-edge technology being efficiently (in terms of both time and financing) developed in new small biotechnology firms, rather than through large corporate R&D groups that partner with small companies to develop new technology. In this probable scenario the effective development of tomorrow's products through new technology is more likely through partnering than large company investment in core technology.

˘ The smaller companies' interest in partnering will be less for the purpose of obtaining capital (although that will always be useful), than for synergy in developing the technology and downstream access to regulatory support, thus defraying some costs of the clinical trials and withstanding potential regulatory delays. Also, partnering will be a vehicle to achieve some form of integration (i.e., manufacturing, marketing, distribution, support).

˘ Large U.S. and European companies will increasingly be the partners in these early stages (due to compatibility of cultures, markets, and comfort with partnering), with Japanese companies involved in later stages. When access to Japanese regulatory processes and Japanese Asian markets becomes more important, partnering with Japan will begin. Since the time frames to complete Japanese linkages are often longer, and thus more costly to small U.S.-based biotechnology firms, the probability of large numbers of U.S.-Japanese strategic alliances in these earlier stages is low.

U.S. biotechnology firms are generally developing a new structural paradigm. Rather than emerge as an industry generating a lot of fully integrated pharmaceutical/chemical/agricultural or food-type companies, only a few (less than 3 percent) fully integrated companies are likely to emerge. Most will integrate in niches but be more strategically linked and partnered, thereby building value and survivability but not growing up to look like Merck or Monsanto. Only a handful of large fully integrated companies will emerge. In the past 50 years, only one major U.S.-based pharmaceuticals company with sales in excess of $1 billion_Syntex_has emerged in this way. As an industry, the biotechnology industry will spawn a few more Syntex's but not many.

Hundreds of strategically linked and niched companies *will* spawn and survive. (These firms will be self-sustaining and vertically integrated in some markets or geographical regions, but will be strategically linked to large foreign and domestic partners in other areas.) The U.S. government can play an active role in this process through mechanisms such as support for industry-driven R&D institutes that focus on areas of generic technology necessary for future applications.

Large U.S. companies in pharmaceuticals, chemicals, energy, and food will increasingly look to the biotechnology world as a vehicle to integrate new technology development. Investments in emerging technologies and early-stage development of products (even through venture capital funds) will therefore be outsourced from the biotechnology world, rather than through capital investment in corporate R&D.

Japanese companies will increasingly pay attention to the U.S. biotechnology industry. However, U.S. biotechnology firms' preferred linkage partners will be European and U.S.-based companies because of cultural compatibility and efficiency. Over time a model will emerge in which Japanese companies will come to parity with U.S. and European companies as potential partners for biotechnology firms, but this will occur over a longer period of time (toward the end of the 1990s). Selectively, we will see acquisition of biotechnology firms by foreign-based investors, with the Japanese companies lagging behind both the European and the American companies as acquirers.

On the other end of the spectrum, Japanese companies are increasing, and will continue to increase, their investment in technology directly with the U.S. academic sector, in lieu of (or in addition to) linkages with small firms. As academic institutions around the world, and principally in the United States, look for corporate support for continued funding, Japanese companies are likely to become increasingly involved. Only very specific technology and product-related links are likely to be the focus of strategic linkages between Japanese companies and U.S.-based biotechnology firms.

Japan will continue to increase its investment in the basic research sciences at a slow rate. The gap among the U.S., European, and Japanese science bases will diminish as we approach the year 2000, but the United States will continue to dominate in the 1990s. It is unlikely that the Japanese companies will attract large numbers of U.S. scientists to move to Japan. Therefore, the trend will be for Japanese companies to build research facilities in the United States as a vehicle to develop and export technology and completed products to Japanese companies for worldwide commercialization.

Since the worldwide pharmaceuticals industry is not dominated by Japanese companies, nonhealth care biotechnology (i.e., electronics, energy, chemicals) is more likely to be developed by them first. Conversely, healthcare-related biotechnology will more likely be developed and partnered by U.S. and European companies (a world in which they dominate).

U.S.-based biotechnology firms are not likely to have the resources to establish the necessary manufacturing, marketing, and sales infrastructure to support the development of their Japanese operations. Therefore, as U.S.-based industry commercializes its products, it will use Japanese linkages for the purpose of establishing marketing, manufacturing, and sales presence in Japan.

With respect to intellectual property rights, differences between Japan's first-to-file system and the U.S. first-to-invent system will cause U.S. biotechnology firms to continue to establish their own intellectual property protection in the United States and Europe and to look to Japanese companies as their partners to establish intellectual property protection in Japan.[72] Many CEOs of U.S. biotechnology firms perceive a gap between U.S. and Japanese intellectual property protection to be a major hindrance to their independent participation in the Japanese market. In other words, U.S. biotechnology firms will establish their own regulatory interfaces with U.S. authorities (or partner with large U.S.-based companies) and will establish their own European-based regulatory interfaces (or partner with European-based companies). In Japan, in contrast, U.S. firms will not establish their own regulatory interfaces but will partner with Japanese companies to facilitate their interfaces. In some cases the U.S.-based biotechnology firms will partner with U.S. and European companies for worldwide regulatory interfaces, with special linkages to Japanese firms for the Japanese market.

Historically, biotechnology R&D has been developed largely in the United States and to a lesser extent in Europe. As we move through the 1990s, Europe will move quickly but will not achieve parity with the United States. As the end of the 1990s, Japan will become increasingly important as a source of technology development. In this regard, U.S.-based biotechnolo-

---

[72]For a detailed treatment of these questions, see Committee on Japan, National Research Council, *Intellectual Property Rights and U.S.-Japan Competition in Biotechnology*, op. cit.

gy firms will begin to partner with Japanese companies for access to new intellectual capital by the end of the decade.

It is likely that the future scenario will reflect U.S.-based development of health-care-related technologies and products with partnering with Japanese companies for access to Japanese markets. At the same time, non-health care technology can be expected to develop more quickly in Japan and U.S.-based companies will seek strategic alignment with Japanese companies to access these technologies for the U.S. and worldwide markets.

To understand any scenario for the future of the U.S. biotechnology industry, one needs to be reminded of the intimate linkage the industry has had to date with a capital market, a market that has generally been very supportive of biotechnology firms. Should this change in the United States during the 1990s, there will be more Japanese linkages as sources of capital, and, therefore, more strategic alignments.

The overall conclusion drawn by the NRC working group is that current trends will continue but that there may be some new developments in Japan that bear careful analysis and response. U.S. industry has some very important assets, but these must be managed in order to ensure that it remains strongly competitive in global biotechnology markets.

5

Conclusions

Various studies, scholars, economists, and politicians have argued, and the NRC working group agrees, that America's ultimate comparative advantage lies in its ability to develop and use technology, because this ability is a major driving force for continued economic growth. Venture-capital-driven investments have played a major role in the development of new technologies and are uniquely American.

The Council on Competitiveness, in a recent report entitled *Gaining New Ground*,[73] describes priorities for America's future and makes specific recommendations that are consistent with this report. According to the council's report, the U.S. position in biotechnology is strong, and the United States is not in danger of losing this leadership in the next 5 years. The NRC working group agrees but notes that there are a number of areas, such as fermentation, scale-up technologies, biosensors, and agricultural applications, where Japanese capabilities approach those of the United States. We are also concerned about the 5- to 10-year window. The present report deals primarily with U.S.-Japan technology linkages and their long-term significance to the United States, a subject not covered in the council's report. This analysis leads us to conclude that *the U.S. biotechnology industry will lose its strong leadership position in several industry segments at the end of the decade unless concrete steps are taken by government, industry, and universities.*

---

[73]Council on Competitiveness, *Gaining New Ground: Technology Priorities for America's Future*, 1991.

The U.S. biotechnology industry is growing very rapidly. In 1990 it had sales of $2 billion. For the year 2000 sales estimates of the U.S. industry vary from $30 billion to more than $70 billion. Biotechnology will be a major factor in the health care industry, even though sales of biotechnology products will represent less than 10 percent of the total. By the year 2010, the role of biotechnology in health care will increase substantially. During the 1990s, many more biotechnology companies will be created than the number that go out of business. The total number of companies will continue to increase, particularly in the health care area. Consolidation of the health care industry will be very slow.

This dynamic industry has emerged from the world's premier science and technology base in the United States and America's venture capital/public market system. In order for the U.S. biotechnology industry to remain strong, however, it is not enough to do good R&D and create research companies. A new generation of scientists and engineers must be trained. Heavy capital investments in manufacturing technology, product improvements, and global marketing of products also are essential. The new ideas created in the laboratory must be taken to the market, and the market is more and more a global one.

A primary conclusion of the NRC working group is that cooperation between the United States and Japan in biotechnology is inevitable and desirable as we move toward a global market, but policy must be developed. The United States and Japan have different resources to bring to bear in developing biotechnology, as a result of striking differences in the two countries' research systems, industrial structures, regulatory regimes, patterns of disease and food products, drivers of capital market systems, and customer access and delivery. To achieve the potential gains from these differences, however, U.S. firms will need to develop careful strategies that focus linkages with Japanese firms on areas where technology transfer is feasible, such as bioprocessing. U.S. firms that want to do business in the Japanese market and tap into Japanese technology will need to develop effective ways of operating in Japan. Increased cooperation among U.S. companies also is desirable.

There are good reasons to suggest that special efforts will be needed by U.S. government, industry, and academe if the benefits of cooperation with Japan are to be maximized. *The danger is that, if conscious strategies are not developed by the U.S. participants to increase inflows of technology from Japan and to expand marketing and sales in Japan, the net result of increasing technology linkages in biotechnology will be to create significant competition from Japan without strengthening the ability of U.S. firms to compete and commercialize technologies.* There are powerful forces driving small innovative U.S. biotechnology firms toward relationships with large capital-rich Japanese firms. These forces include the need for infusions of capital to ensure survival today; they are unlikely to weaken in the decade ahead. The strength of the venture capital industry and the ability of U.S. biotechnology companies to gain access to venture capital are important factors that make it more likely that linkages will work to strengthen the U.S. competitive position.

The price run-up of biotechnology stocks and the large amount of new equity capital raised by the industry during 1991 have alleviated the immediate financial needs of some top-tier companies. However, the financing issue will not recede in importance as a result of the boom in initial public offerings (IPOs) and secondary offerings. Several factors should be kept in mind. First, the benefits of access to public markets have accrued to a relatively small number of firms in the top tiers. Second, public markets cannot substitute for seed capital _ the amount available for venture capital investment in biotechnology has declined in recent years along with the overall pool of venture capital. Third, even for the companies with access to public markets today, the favorable climate will not continue indefinitely, and the resources required to bring products through the regulatory process to market remain considerable. In short, financing remains an immediate concern for the vast majority of the U.S. biotechnology industry, and remains an underlying issue for even the top companies.

To assess the impacts of U.S.-Japan technology linkages, it is important to evaluate the effects of various types of linkages by considering a range of possible impacts. The possible effects include capital generation, expanded market access, profits, number of jobs, creation of new technology, expanded R&D, and effects on the broader national scientific manpower base. Detailed analysis of particular cases also is required. From a U.S. perspective, linkages that result in a weakening of downstream activities (manufacturing, marketing, sales) in the United States raise significant concerns. On the other hand, U.S.-Japan linkages that strengthen the U.S. research base and accelerate commercialization can bring real benefits. The reality is that the short- and long-term impacts may differ significantly and that technology linkages normally involve multiple mechanisms. Research conducted by the NRC working group indicates that U.S.-Japan linkages established to date have overall had positive effects in infusing new capital and expanding the R&D efforts of U.S. firms. Technology transfers to Japan have not resulted in an erosion of the capabilities of U.S. firms in biotechnology product sales. However, during the past decade, Japanese industry, which views biotechnology as the most promising technology for future growth, has positioned itself well to compete, particularly in markets outside the health care area and particularly in Asia.

The NRC working group's concern is with the future rather than the past. The technology linkages that have already been formed, particularly those involving licensing of technology, may have significant market impacts in the decade ahead. If the pattern in the future continues to be a one-way technology transfer to Japan without the development of strong global commercialization capabilities in the United States, the results could be significant and negative by the end of the decade. Japan is well positioned to fuse biotechnology with other technologies since most of the biotechnology products are being developed by large companies in the food, chemical, and health care industries.

A key question is whether U.S. organizations involved in biotechnology_including universities and nonprofit research organizations_can make the technology linkages with Japan work to their long-term benefit. To do so, a strategy must be developed to ensure that R&D capabilities in the United States are maintained and strengthened; that technology continues to be transferred out of U.S. universities; that the intellectual property rights of innovators are protected; that technology transfer from Japan in areas such as microbial cell lines, bioprocessing and automated screening assay techniques, and biosensor technology takes place; and that U.S. organizations find expanded opportunities to participate in the Japanese regulatory process and in the Japanese market and are able to enforce patent rights. To assess the long-term significance of technology linkages, it is important to step back and see the trends and the potential significance for the United States as a country as well as for the organizations directly participating.

*Despite the strengths of the U.S. biotechnology industry today, the NRC working group is not sanguine about the future and the ability of the U.S. biotechnology industry to compete in the twenty-first century. Significant potential problems were identified that cannot be adequately addressed on an ad hoc basis because active collaboration of government, industry, and universities will be needed*_perhaps through an industrial and technology policy for biotechnology. The President's Council on Competitiveness recently released a report that is a good starting point. Its major recommendations are consistent with the present report, which focuses more specifically on U.S.-Japan linkages, technology transfer, and the importance of emerging firms.

The NRC working group believes that expanded cooperation among industry, academe, and government is needed, particularly outside the health care arena where serious competition will emerge first. The subsequent sections of this chapter identify specific issues that each sector must address, but in most cases effective response will depend on joint efforts.

ISSUES FOR INDUSTRY

The U.S. biotechnology industry must focus more resources on product development, production technology, and marketing of new products globally while at the same time maintaining excellence in research. To do this, *it will be necessary for U.S. companies to focus investments on second-generation products and improvements (vs. generating breakthroughs)*_an area of strength for Japanese companies. A new partnership between industry and government to establish technologically oriented institutes may be essential.

Fusion of biotechnology with other technologies such as electronics, purification of chemicals, applications in agriculture, and food must be pursued through technology linkages. Continuous improvement and refinement of products and technology are a prerequisite for staying on the front edge of the competitiveness curve. This strategy has major implications in the regulatory field and in the funding of new technology. (See Issues for Government below.)

The U.S. biotechnology industry must seek expanded participation in commercialization and direct marketing of its products around the world, particularly in Europe and Asia. An exclusive reliance by U.S. firms on licensing out first-generation products, marketing, and sales rights ultimately creates increased competition. U.S. companies, particularly small biotechnology firms, must participate in the global commercialization of products. Comarketing and joint sales by U.S. and Japanese companies should be preferred over exclusive licensing out of products in return for short-term infusions of capital.

The U.S. biotechnology industry must develop a global strategy, and Japan, as the world's second-largest pharmaceuticals market, and the European Community must be key elements of that strategy. This means building the expertise to play in the Japanese market_to carry out clinical tests, to market products, and ultimately to tap into Japanese R&D. Unfortunately, the cost of establishing a business in Japan is prohibitive for many small U.S. biotechnology firms.

U.S. biotechnology firms will need to develop new approaches to structuring linkages to become more global. One example would be a sequential approach in technical licensing that ensures, over time, comarketing and market access opportunities in Japan for products from emerging companies. Another possibility would be to develop linkages that combine capital and technology inflows from Japan. Still another possibility would be to develop improved collaboration within U.S. industry, perhaps through joint efforts to keep track of the policy and market contexts in Japan by industry associations or *more creative agreements among small U.S. companies and larger U.S. companies* that allow the smaller companies to begin product manufacturing within 5 to 10 years. In other words, small U.S. companies could rent capabilities in the larger U.S. companies versus selling all future rights abroad. If such approaches are to work, the large U.S. companies must play a more significant and aggressive role in building strategy and in focusing on biotechnology-related applications.[74]

---

[74]One example of a linkage between a large U.S. pharmaceuticals company and a small biotechnology firm is the acquisition by American Home Products of 60 percent of Genetics Institute in the fall of 1991.

Thus, for industry, the key issues are (1) *the need for greater investments in product and technology improvements, (2) the importance of global commercialization of products, and (3) the need for increased financial staying power for the emerging companies.* These require financial strength, and government can help. (See Issues for Government below.) In other words, efforts to improve capital formation are a prerequisite to addressing these issues. The American university/venture capital industry relationship is essential for long-term competitiveness in the years ahead.

ISSUES FOR GOVERNMENT

In view of the critical importance of investment to the health and viability of the U.S. biotechnology industry and other R&D-intensive industries, *the U.S. government could increase financial incentives to encourage innovation, more venture and patient capital, and long-term strategy building.* Technology development is an area of current U.S. strength but is also an area where renewed efforts will be important to ensuring future competitive strength. New approaches should be considered that feature government and industry working together to develop generic technologies (precompetitive) important to future applications. This is an area where the United States can learn from studying Japanese approaches.

The NRC working group considers investment-related measures that lead to easier capital market access and greater financial strength for emerging companies to be top priorities. Possible approaches that require further study and debate include making the R&D tax credit permanent, introducing a graduated capital gains tax for technology investments, and establishing a pool of patient capital to provide seed investments for promising innovators.

U.S. government policy in the regulatory and international trade spheres also has a direct impact on the financial strength of innovating firms, particularly biotechnology firms. The provision of resources to speed the review process for new biotechnology products while maintaining the highest safety and efficacy standards is one possible focus. In the trade area, the possiblity of initiatives by the Agency for International Development (AID) and the International Trade Administration to promote exports of U.S. biotechnology products is worth exploring.

The U.S. government could consider developing, in cooperation with industry, a technology strategy for biotechnology as a scientific enterprise and technology, giving special attention to the unique contributions made by small entrepreneurial firms and U.S. research universities. A government-industry forum could be established to enhance discussion and debate. Agencies funding research could place a special priority on non-medical applications, perhaps working with state biotechnology centers.

The U.S. government should continue to evaluate the creation of centers of excellence that act as bridges between universities and industry.

These centers need long-term funding from government and industry. They could contribute to the U.S. biotechnology industry by carrying out research programs that focus on enabling technologies and technology fusions that benefit many companies and industries. Since biotechnology is a new set of tools that could be used by a variety of industries, there is a rationale for developing a plan of broad-based support for research, development, and training. Participation in such programs should be permitted for foreign companies that are manufacturing and performing R&D in the United States and whose home governments provide similar opportunities to U.S. firms.

The U.S. government could also encourage the appropriate agencies to define programs that have as their purpose increasing information about new developments in Japanese biotechnology R&D to U.S. industry. For example, a program of cooperation with a Japan industry-based biotechnology R&D consortium could feature on-line access for U.S. researchers to Japanese biotechnology data bases, patent registrations, and electronic mail reports on new Japanese research projects in biotechnology with information about themes, participants, and laboratories. In order for such efforts to be meaningful, however, a Japanese-speaking U.S. researcher should be present in Japan to participate in identifying useful information and to communicate with users in the U.S. biotechnology research community on a regular basis.[75] Joint projects involving companies and universities from both countries could also be developed that have as their goal incremental technology development.

In addition to programs aimed at gathering and disseminating technical information, careful assessments by U.S. government and industry of differences in accounting practices, regulatory environments, and government support to biotechnology R&D in the United States and Japan could form the basis for informed discussions with Japanese counterparts.[76] These efforts would also support the strategy-building by U.S. government and industry mentioned above.

The U.S. government could give serious consideration to a move to the first-to-file system of intellectual property rights protection. The current system gives strong incentives to innovators, thereby supporting the work of small firms. At the same time, however, serious differences exist between approaches to intellectual property in the United States and Japan.[77]

---

[75] See Committee on Japan, National Research Council, *Expanding Access to Precompetitive Research in the United States and Japan: Biotechnology and Optoelectronics* (Washington, D.C.: National Academy Press, 1990).

[76] The U.S.-Japan Business Council established a working group on biotechnology that has recommended regulatory changes in both countries. The patents offices of the two countries also are working together to promote the expeditious publication of patent data in English as well as Japanese.

[77] See Committee on Japan, National Research Council, *Intellectual Property Rights and U.S.-Japan Competition in Biotechnology: Report of a Workshop* (Washington, D.C.: 1991).

Harmonization around a first-to-file principle would bring U.S. practices in line with those of other major competitors and would enhance the ability of entrepreneurs in U.S. industry, national laboratories, and academe to profit from their innovations in Japan and around the world. Domestic policy giving strong protection to biotechnology processes as well as product patents and international harmonization around similar principles will also benefit U.S. industry.

ISSUES FOR UNIVERSITIES AND NONPROFIT RESEARCH INSTITUTIONS

Cooperation with industry must be deepened, and *technology transfer should be supported as an important activity*. To make the most of collaboration with industry while ensuring academic freedom and proper protection of intellectual property rights, universities should develop standard policies for licensing technology to industry that permit faculty participation in the formation of new companies. Because funding of research and participation by Japanese companies are growing, *universities could develop guidelines for good practice for contract research, standards for the conduct of foreign researchers in U.S. laboratories, and other measures to ensure reciprocal access* for researchers to the laboratories and know-how of the foreign sponsoring organizations. Such measures may include laboratory access in Japan, cross-licensing rights, and access to improvements in technology originally developed in the U.S. university setting. Dedicated centers of excellence that are industry driven should be considered as a means to provide longer-term orientation to the development of new technologies.

U.S. universities must maintain their excellent research base, but work to develop global thinkers and entrepreneurial managers in all disciplines. Programs of Japanese-language training and professional experience in Japan for scientists and engineers should be expanded. Part of the science curriculum should include the fundamentals of international business. In addition, cross-fertilization should be promoted among technical researchers, business strategists, and area specialists through professional associations and university courses.

Professional associations may find it necessary to develop strategies for international participation. In this regard, *cooperative efforts to collect and analyze data on foreign participation* in U.S. university research could make a significant contribution to better public understanding of reciprocity issues. Professional associations can also encourage Japanese-language study for members through fellowship programs and information dissemination.

U.S. universities and regional biotechnology centers may also contribute to public education about biotechnology. Biotechnology centers are training high school biology teachers and developing curricula that feature hands-on laboratory experiences. There is a stronger emphasis on public education in Japan, and the companies, through their advertising campaigns, reinforce a positive image for biotechnology-based products.

This is the right time to look ahead to the year 2000 and the future of U.S. biotechnology_as a research and market enterprise. The technology linkages that are being formed with Japan present opportunities as well as risks for the U.S. partners_companies, universities, and national research institutes. For a variety of reasons examined in this report, there is little question that these linkages will expand and deepen. The question that must be asked is whether in the year 2000 it will be clear that they have produced concrete benefits to both the United States and Japan. The answer depends, to a great extent, on whether U.S. organizations can individually and cooperatively develop new ways of interacting with each other and Japan and whether the American innovation machine will continue to be a leader in the development of new technologies and products.

Appendix A

Case Studies of U.S.-Japan Technology Linkages in Biotechnology

CASE I: CALGENE-KIRIN[78]

An agreement was reached in March 1990 between Calgene and Kirin to jointly develop and market potato seedlings. Although the partnership is still relatively young, which makes it difficult to assess its impacts on the two companies, the joint venture illustrates some of the business issues that are relevant to agricultural biotechnology. In addition, the alliance contains several novel structural features that may shed light on possible future directions for U.S.-Japan biotechnology linkages.

The Partners

It might be useful to begin with a description of the partners and where agricultural biotechnology fits into their businesses.

Kirin is the fourth-largest brewer in the world, with unconsolidated sales of about $10 billion. The nonbeer businesses that contribute significant amounts to sales include engineering services (centering on bottling factories), food, and soft drinks.

---

[78]Subsequent to the preparation of this case study, Calgene announced that it was restructuring as a result of "significant breakthroughs in...core crop areas." As a part of this restructuring, the joint venture with Kirin was to be downsized. See "Calgene Restructures Operating Businesses to Focus on Three Core Crops," *Biotech Patent News*, September 1991, p. 5.

As a result of the long-term vision adopted in 1980, which put forth the goal of becoming a company that "contributes to life and health around the world," Kirin began to diversify into services (restaurants); engineering; information services; food products (dairy, tomatoes, coffee); and life sciences (pharmaceuticals, agricultural biotechnology). Corporate R&D spending is $110 million per year. The firm's principal subsidiaries are the Kirin-Seagrams joint venture, Coca-Cola bottling franchises in western Japan and New England, and the Kirin-Amgen joint venture to manufacture and market EPO and G-CSF (see Case III below). Kirin has more than 30 domestic and overseas subsidiaries.

The focus of Kirin's Agribio Division is the plant laboratory. Kirin is seeking to utilize biotechnology to develop new varieties for mass propagation. To compete with established seed companies, a new strategy was adopted that incorporates the use of cell fusion and artificial seed technology for breeding and propagation, an emphasis on "seedlings" rather than seeds, leveraging the strong brand consciousness of Kirin products, and formation of a global network of subsidiaries and joint ventures. Globalization makes it possible to exploit market opportunities quickly. Joint ventures with companies possessing complementary technologies are particularly attractive because they allow Kirin to maximize the return on technology developed internally. Kirin's other partnerships in agricultural biotechnology include Tokita Seed (vegetables), Flower Gate, and Twyford (in vitro plants).

Calgene, founded in 1980, is a publicly traded company that focuses exclusively on agricultural biotechnology. The firm's projected revenues for 1991 were $35 million; it has spent a total of $70 million on R&D; and it has raised $120 million to $130 million in capital since its founding. It was the first company to apply for Food and Drug Administration (FDA) approval of genes to be introduced into plants. The firm's core products are genetically engineered tomatoes, cotton, and rapeseed. Calgene has 300 employees, including 100 scientists, in five operating groups. Half the employees are located at its headquarters in Davis, California.

Calgene is actively pursuing vertical integration, seeking direct access to markets in all of its core businesses.

Origins of the Linkage

Long-established business and personal relationships as well as a "strategic fit" were crucial in putting the partnership together. In 1984 Kirin bought an equity stake in Plant Genetics, Inc. (PGI), the agricultural biotechnology company that later merged with Calgene. PGI also performed contract research for Kirin in the area of synthetic seeds. Zachary Wochok, a founder of PGI, worked with Yoshihiro Imaeda and Kirin's legal representative, Joel Marcus, to set up the alliance. Both companies were doing research on potatoes but did not cooperate in this area from the beginning.

Several factors make the potato market an attractive target for the application of agricultural biotechnology. First, potatoes are a multibillion dollar crop worldwide. Consumption in the United States is growing at a rate of 7 percent per year, mainly due to sales of french fries and other fast foods. Second, potatoes are relatively easy to manipulate through genetic engineering. Finally, although governments around the world are involved in trying to improve potato yields and quality, there is very little private sector involvement or market discipline.

One key technical issue is reduction of the "bulk-up" period required for seed potatoes. Potatoes are grown from "seed pieces," and it takes 7 years, using conventional techniques, to produce enough seeds to sell to farmers, a process known as bulk-up. Reduction of the bulk-up period to 3 to 4 years would result in a significant efficiency gain. If the period could be reduced to 1 to 2 years, the resulting proprietary product would drive the potato market.

Kirin first approached PGI about extending its collaboration to seed potatoes after the latter's initial public offering in 1987. PGI was investigating the introduction of genes into potato varieties to promote pest and disease resistance. For its part, Kirin had developed a technique, called the "microtuber," that allows generation of a seedling from a single cell rather than through seed pieces. If it performs up to its potential, the technology will allow a reduction in the bulk-up cycle to 1 to 2 years. Wochok and others at PGI were skeptical at first, but Kirin continued to update them on their progress.

Negotiation Process and Issues

In 1989 PGI and Calgene, located next door, merged. Kirin's PGI stock became Calgene stock, and Kirin again raised the question of collaboration in potatoes. Discussion continued through 1989. That year Calgene researchers went to Hokkaido for a Kirin presentation on its microtuber field experiments. Calgene became more confident, although questions remained about scaling up the technique and how effective it would be in the United States. Calgene had already achieved a reduction of the bulk-up period to the 3- to 4-year time frame through its own ongoing research program and was selling pest-resistant seed potatoes to farmers.

The main issues were the valuations of Calgene's seed potato business and Kirin's microtuber technology. The former issue was the main stumbling block. The valuations of Calgene's potato receivables, inventory, and other assets made by the two sides at the start of discussions were disparate by a factor of 10. The two sides resolved their differences on this point during a day-long meeting in December 1989, which was a "make-or-break" session for realizing the partnership. Kirin used stock price valuation and projected profit/price earnings ratio figures submitted by Calgene. Kirin also calculated new figures based on its own assumptions about growth prospects and factored in what it would contribute to the venture.

At the end of 1989 the companies shook hands on the basic agreement. The formal negotiations were completed less than 3 months later without the involvement of investment bankers. The basic outline was for Kirin to provide financing to the venture and for Calgene to contribute the personnel and core technology. Kirin already had a high opinion of PGI (now Calgene) personnel, quality control practices, and the systematic collection of germ plasm. In addition, Kirin determined that a partnership with Calgene was the most effective way to commercialize its microtuber technique. Calgene's experience with recombinant DNA and cell fusion in many species and its collection of genes isolated for possible introduction into potatoes were additional benefits that a premerger partnership with PGI would not have provided. Ideally, the venture will be able to introduce genetic improvements into popular types of potatoes and provide a new, more efficient production method.

Structure of the Linkage

The Kirin-Calgene partnership contains a number of elements, including equity, licensing, and contract research. Kirin made an initial asset purchase of 30 percent of Calgene's seed potato business for $2.5 million. The companies formed an operating joint venture, called Plant Genetics-Kirin (PGK), in which Calgene held a 70 percent share and Kirin 30 percent. Kirin has since increased its stake to 35 percent. Kirin licensed its production technology_the microtuber technique_to PGK and will be paid in a series of "equity kickers." If the microtuber reaches "agreed performance milestones," Kirin's stake in the joint venture will rise to a maximum of 50 percent. Calgene now appoints two members and Kirin one to PGK's management committee. When its equity reaches 40 percent, Kirin will add a financial representative to the committee.

In evaluating Kirin's technology, the idea of "equity kickers" was suggested at an early stage. The basic concept is that, to the extent that Kirin's technology works, the value of PGK will increase and so should Kirin's stake. The performance milestones are qualitative rather than quantitative, and both sides are confident that they will be able to agree on whether they have been met. This structure protects Calgene and rewards Kirin if its confidence in the microtuber technique is justified. The phased growth in equity stakes also addresses Calgene's reluctance to go into a 50-50 joint venture at the outset.

In another set of contracts, Calgene licensed its own core technology to the venture. This consists of genes that are introduced into potato varieties to make them more resistant to pests, particularly the Colorado potato beetle. Outside of the PGK framework, Kirin is paying Calgene $1.5 million over 2 years for contract research on potato genes that promote pest and disease resistance. Kirin is the Asian licensee for Calgene's core pest-resistance technology, and the joint venture is the licensee for the rest of the world.

The agreement also contains termination mechanisms. Corporate strategic objectives may change over time, but the joint venture has explicit goals built into the business plan. The companies hope that the management systems put in place will ensure that the objectives_both annual and over the 5-year horizon_will remain explicit and are incorporated into the strategic planning of each side. Budgets and other operational matters will remain manageable if both companies remain focused on clear strategic goals.

Strategic Goals, Management, and Technology Transfer

Calgene faced a basic dilemma of how to pursue new opportunities, in areas like potatoes and alfalfa, while pushing for vertical integration in its tomato, cotton, and rapeseed products. The distribution system for potato seedlings is fairly complicated. At this point, PGK sells seedlings to farmers, but the key to future profitability will be the degree of vertical integration that can be achieved. With the combination of better yields as a result of the pest-resistance features and a shortened bulk-up period, the venture's superior product may allow it to move downstream. Ideally, rather than selling a "turnkey" product to farmers, PGK would contract with them, process the crop, and then negotiate directly with major consumers like McDonald's.

Besides the contribution of complementary technology, Kirin's participation also ensures that the resources for a worldwide push, particularly into the critical European market, will be available as products come on line. Although Calgene has a joint venture in Scotland, it would be very difficult for the firm to move quickly into foreign markets by itself. Kirin's clear commitment to potato development was another factor that made it an attractive partner for Calgene.

PGK itself has a marketing and sales emphasis_intellectual property rights to technologies are retained by the parties and licensed to the partnership. At this point, Calgene charges the venture for facilities and personnel, but PGK itself was expected to begin hiring its own employees in late 1991. Technical exchange goes on between the two partners through reciprocal research exchange visits and placements of up to 3 months. As in most biotechnology linkages that involve researcher exchange, the mechanisms are written into the agreement. So are the monitoring and control systems. These center on quarterly meetings of scientific counterparts, business counterparts, and a joint science-business session.

Kirin is responsible for producing a quality microtuber efficiently and PGK will be responsible for field testing in the United States. Calgene sent researchers to Hokkaido during 1990 to get the benefit of Kirin's experience in running field tests of microtuber potatoes. Even though the American personnel are responsible for meeting the performance milestones, Kirin can visit at any time to evaluate the field tests.

Whose technology is more critical to the venture's success_Calgene's or Kirin's? The answer is still uncertain. Clearly, Calgene's pest-resistance genes are the basis of the current commercial effort, but the performance of the microtuber technique will directly impact the degree to which PGK can vertically integrate. Ultimately, this will determine PGK's profit margins.

Even though the initial dollar amounts of the partnership's various elements are small, the potential importance of the product and the belief that "informality does not bind" led both sides to conclude that a structured joint venture would be more efficient in the long run than an informal collaborative arrangement.

What are PGK's weaknesses? One vulnerability that often arises in U.S.-Japan biotechnology linkages is overdependence on the contributions of particular individuals in making the alliance a success. It is often the case that long-standing relationships facilitate the formation of a venture, but this also means that partnerships rely heavily on the key players to keep the business on track and to resolve disagreements. Since personnel rotation and lifetime employment are still standard human resource management practices in large Japanese companies, the problem of vulnerability is more likely to arise on the U.S. side. In the case of PGK, Zachary Wochok played a key role in building the alliance. As deeper relationships are developed between the scientific and business sides of the partners and PGK develops its own momentum, PGK will be less dependent on the contributions of key individuals.

Those involved in putting PGK together cite several key elements that allowed the two sides to come to an agreement. One was the strategic fit of complementary technologies and capabilities. Previous relationships also were important. Wochok and Joel Marcus played key roles. Though the latter serves as Kirin's legal representative, Calgene had confidence in him because of a previous association. Another important element that contributed to forming the venture was the equity enhancement mechanism. Finally, patience, determination, and regular face-to-face communication during the negotiating process also were critical.

CASE II: MONOTECH, INC. AND SHOWA-TOYO DIAGNOSTICS[79]

The distribution and licensing agreement between Monotech, Inc. and Showa-Toyo Diagnostics (STD) covering Monotech's in vitro cancer diagnostic products and technology was concluded in 1982, and the first products were on the market in Japan in 1985. Monotech is a rapidly growing U.S. biotechnology firm, and STD is a joint venture between a large Japanese textile and chemical manufacturer and a medium-sized health care company. The arrangement currently covers five products, and the venture has annual sales of 2.1 billion yen ($15.5 million at 135 yen per dollar).

Because the relationship has a substantial track record, it is possible to look back and assess the effects on the firms. Other emerging U.S. biotechnology companies may be able to learn from Monotech's experience. Looking to the future, it is also possible to ask whether changes can be made in the structure of the linkage to ensure that it serves the strategic interests of the partners in the 1990s as well as it did during the 1980s.

The Partners

To fully understand the role of the linkage in the strategies of the partners, it is necessary to begin with a brief overview of the companies and the role of biotechnology and diagnostic products in their businesses.

Monotech is one of the leading U.S. biotechnology companies. Its research and market focus is on the field of monoclonal antibodies. The in vitro diagnostic products that are the basis of the linkage to STD were Monotech's first commercial products, and income from them has played a major role in bridging the gap to the revenue stream expected from the company's first therapeutic product, Mabex. Mabex is a treatment for gram-negative sepsis. In 1990 Monotech registered sales of over $30 million and comparable income from R&D limited partnerships. The company registered a net loss of about $130 million as a result of exercising options to buy back shares in several of the limited partnerships it had set up to fund product development. Monotech spent about $45 million on R&D in 1990 and had almost 900 employees as of March 1991.

Showa Materials is one of Japan's leading textile firms, specializing in synthetic fibers. It has diversified aggressively into engineering plastics, carbon fibers, and health care, and about 10 percent of its sales are in "new operations." In the health care area, in addition to its cancer diagnostics

---

[79] The names of the companies, products, and individuals that appear in Case II have been changed.

activities through STD, Showa has developed antiulcer and cardiovascular therapeutics. In biotechnology Showa has several marketing agreements with U.S. companies for diagnostic products, and in therapeutics the company has put its emphasis on beta interferon, where it has collaborated with both American and Japanese companies. Showa earned a net profit of 40 billion yen ($296 million at 135 yen per dollar) on consolidated sales of 844 billion yen ($6.2 billion) in the fiscal year ending March 1990. Showa Materials has over 10,000 employees and owns about 6.1 percent of Toyo Health.

Toyo Health is a health care company specializing in clinical reagents. In 1989 it earned a net 1.6 billion yen ($12 million) on consolidated sales of about 60 billion yen ($440 million). Besides the cancer diagnostics based on Monotech's analytes, Toyo distributes and manufactures other human diagnostic products under license. Toyo has over 800 employees, a figure that does not include subsidiaries.

The Products

Monotech's In Vitro Diagnostics Division, headed by Joseph Atkins, is the most established of its business divisions. Its main line of products are cancer blood tests. Monotech sells complete test kits that utilize radioimmunoassay (RIA) methods through distributors as well as analyte (monoclonal antibodies), which is fabricated into kits by several licensed partners. Monotech analytes account for 25 percent of the world end product market for cancer immunodiagnostic tests. Over three-quarters of the Diagnostics Division's sales_which have represented the bulk of Monotech's total sales to date_are international, and one-third of the international sales are in Japan.

A monoclonal antibody clings to a single antigen and is produced by a single B lymphocyte. In the mid-1970s biochemists developed a method to capture individual antibodies and the cells that produce them. Among the applications of monoclonal antibodies to health care is in immunoassays to detect tumor antigens secreted into the blood. A blood sample is combined with the analyte and its chemical tag. The antibody binds up with the antigen, and the amount of antigen is then measured by comparing the sample profile to a reference curve.

Monotech's five main products in the in vitro diagnostics field are:

1. MI-1 is used to detect ovarian cancer. It is approved for use in Japan (1986), Europe, and the United States (1987).
2. MI-2 detects and monitors gastrointestinal and pancreatic cancers. It is approved for use in Japan (1985) and in Europe. It is available for experimental use in the United States.
3. MI-3, which is used to detect breast cancer, is approved for use in Japan (1987) and Europe, and is available in the United States for research.

4. MI-4, which detects gastric cancer, is approved for use in Europe and Japan (1987) and is available for research in the United States.

5. MI-5 monitors the resistance of cancer cells to drugs during multidrug chemotherapy. It has been approved for use in Japan and Europe (1989) and is available for investigational use in the United States.

Launching the Partnership

Showa-Toyo Diagnostics is a 50-50 joint venture between Showa Materials and Toyo Health. It was formed for the purpose of marketing Monotech's in vitro products in Japan. During the "biotechnology boom" of the early 1980s in Japan, monoclonal antibodies were one technical area that received a great deal of attention. Showa was already taking steps to diversify into health care and biotechnology and was interested in this field.

During 1982, Monotech's cofounder and current chairman, William Nelson, made extensive efforts to find partners to market the firm's diagnostic products in each of the major markets_the United States, Europe, and Japan. Since Japan has very high rates of gastrointestinal cancer and MI-2 would be the first product to emerge from the pipeline, gaining access to that market received particular emphasis. Nelson visited Japan as part of this effort, and Showa was one of the interested parties. Showa enlisted Toyo for its diagnostics marketing experience. To narrow the field of possible partners to a manageable size Nelson set up a "lottery" for the products. He told interested parties that for a nonrefundable fee of $10,000 they would be considered. A number of companies came forward, and Monotech's management evaluated the business proposals. Through a process that was part intuition and part analytical, the Showa-Toyo joint venture was chosen. In particular, Monotech liked the idea of a separate venture built around the products.

The motivations of the partners were fairly straightforward. Monotech wanted aggressive marketing of its products in Japan. In this field, as in many others, the Japanese distribution system contains layers of wholesalers, and it would be unthinkable for a U.S. company, particularly a start-up company, to contemplate an independent sales effort. Showa wanted experience in the management and marketing of biotechnology products and an opportunity to integrate into manufacturing and development. Toyo wanted Monotech's cancer tests as an addition to its line of diagnostic products.

Structure of the Relationship, Technology Transfer, and Marketing Issues

The linkage between Monotech and STD is fundamentally a licensing and marketing partnership. The contract provides for a transfer of products and assistance in development. Management of the approval process in Japan is the responsibility of STD. Since reactive effects in patients are not an issue for in vitro diagnostics, clinical trials and the approval process in general are not as expensive or time consuming as they are for human therapeutics or in vivo diagnostics. The main issue for establishing a product's effectiveness as a diagnostic tool is an assessment of the level of risk associated with a given blood level of antigen in specified clinical situations.

The products have been very successful in the Japanese market. The gastric cancer rate is high in Japan, and MI-2 is the best way to detect the disease. Monotech manufactures complete kits utilizing radioactive tags that the venture distributes. The venture can also incorporate the analyte into nonradioactive delivery systems. Monotech gets a royalty of approximately 20 percent of end-product sales. About 60 percent of sales are the complete kits that Monotech ships, and 40 percent are royalties on kits manufactured by STD and independently by Toyo. STD itself has over 100 employees, who provide technical support, manage the product approval process, ship the product, and increasingly manufacture and market it. Monotech has two technical meetings each year with STD, one in the United States and one in Japan. Business meetings are held semiannually as well.

The fundamental knowledge that STD requires to support sales of the tests concerns the reactive properties of the analyte. Technology transfer is relatively simple and is accomplished by visits of three or four STD researchers to Monotech for several weeks prior to the technical meetings. In the licensed development of delivery systems by STD and other Monotech partners, Monotech provides more analyte for experimental purposes, and the partners specify generic methods of non-RIA tagging to the particular analyte. Partners that fabricate kits under license also do some purification of the antibody. Monotech does not have its own non-RIA delivery system development program.

In contrast to Mabex and other Monotech therapeutic products, for which large amounts of antibody are required, the manufacturing process for the analyte does not present major problems. Large-scale bioprocessing is not necessary. One gram of antibody will last for a million or so tests, and sufficient amounts can be manufactured easily in mice.

Technology and Strategic Issues

The technology and strategic issues in the field of in vitro diagnostics are somewhat different than those that arise in pharmaceuticals_both therapeutics and pharmaceutical diagnostics. For its therapeutic and injectable diagnostic imaging products, Monotech is trying to build an independent global marketing capability and integrate downstream. For in vitro diagnostics, the company distributes the product initially with a standard radioactive delivery system. Extending the life cycle of the product and reaching a wider market have two interrelated elements.

The first depends on clinical work with the analyte that Monotech undertakes with its research partners all over the world. For example, MI-1 is already used along with other diagnostic methods in detecting ovarian cancer and in "managing" postoperative treatment. Clinicians have more recently found that MI-1 may be cost effective as a screening test for postmenopausal women, since false-positive values of the antigen are less common in older women. In supporting and participating in clinical work on new areas in which the test can be used, Monotech expands the market for the test.

The second element is the delivery system. Monotech depends on its partners to drive the delivery system technology. The MI-2 analyte will likely have a long product life, but the test will be performed differently as delivery systems are improved. One well-known example is the glucose test for diabetes. Chemically, it is the same test that it was 20 years ago, but since then the delivery system has advanced to the point where patients can perform it at home. So Monotech, which licenses kit manufacturing and supplies analyte to several companies on a nonexclusive basis, gets wider breadth from the variety of partners and from competition among those partners. In addition to STD and Toyo, Monotech also licenses delivery system development to a major U.S. pharmaceuticals company and a French health care firm. These licensees sell kits in Japan through different distributors.

To a large extent, the radioactive tag is the classical way of performing immunoassay tests. On a new undefined analyte, it is best to start out with an RIA delivery system. When the performance characteristics are understood and the test is peer reviewed, nonradioactive systems utilizing enzymes (enzyme immunoassay or EIA) and chemiluminescence can be developed. This has many benefits. For example, in Japan radioactive isotopes can be used only in the large reference labs and in hospitals that have special facilities. Many Japanese hospitals cannot perform RIA tests. In addition to the safety issues, an emotional aversion to radioactivity in Japan plays a part. Therefore, access to a large part of the Japanese hospital market requires the development of non-RIA kits and instruments.

In addition to safety and the psychological edge, non-RIA also allows development of kits with a longer shelf life_6 weeks for RIA versus as long as a year for non-RIA. Other "user-friendly" characteristics can be chemically engineered as well. In particular, EIA and chemiluminescence allow a proper reaction to occur even when reagents are added non-sequentially. This means that the test can be automated. The technician, rather than pipetting in various chemicals in a prescribed order, can insert a test tube containing the sample into a machine and walk away, returning later to assess the results.

Development of the Linkage Over Time

Monotech's contract with STD allows the venture to use the analyte in developing alternative delivery systems; indeed, it is in Monotech's interest for the venture to do so. At first, STD wanted to change the contract to allow it to develop RIA systems. This would have transferred kit fabrication to Japan from Monotech but would have done nothing to increase the overall market. One of Toyo's subsidiaries is a reference laboratory, and long-term development of systems to reach the hospital market may not have had the short-term impact on earnings that internalization of RIA kit manufacture would. After intense and drawn-out discussions, Monotech convinced the venture to concentrate on non-RIA kits and instruments.

Since then another complication has arisen. The two parents, Showa and Toyo, are developing competing non-RIA products. Though the venture is 50-50, Showa has more influence on the management of the joint venture. Its program is partially inside STD and is complemented by an in-house effort. Toyo's program is independent. Monotech sells the analyte to Toyo as well as to the joint venture.

Competition between the Japanese parents seems to be taking a toll. Instead of one non-RIA development program in STD, each of the parents has an independent program. There are two system development programs and two kit development programs, and Toyo has a focused monoclonal antibodies program as well. This has escalated the costs of development, resulted in competing products reaching the market, and ultimately led to lower margins. STD has a growing independent sales force as well. It had used the Toyo sales team, but Toyo's development of competing products is making continued use of this channel untenable.

Showa and Toyo now have different ideas about the place of the venture in their strategic visions; both want to integrate into manufacturing and product development independently. This may be diluting their focus on growth and may represent an obstacle to STD becoming a major player in the Japanese diagnostics market in the long term.

Prospects for the Future

Presently, Monotech holds no equity and has no formal management role in STD. It does have influence, albeit distant and infrequent, as the licensor and critical supplier. Up to now, the rapid growth in Japanese sales by the venture has served Monotech well. Will continuation of the current arrangement serve Monotech's interests as well in the 1990s?

An alternative that Monotech might pursue is to seek a more active role in the venture. Monotech's global perspective and insight would likely make a material contribution to increasing the size of the Japanese market for the products. To give the venture closer strategic and tactical direction, it would be necessary for Monotech to let the venture see more of its own technology. Concretely, this would consist of Monotech working more closely with the venture on new analytes and exposing the venture on a more timely basis to clinical work being done. This would give STD an overall marketing edge over other Monotech licensees in the Japanese market and a time advantage in developing delivery systems.

Closer involvement might bring other benefits to Monotech beyond a larger market for its products. For example, Japanese companies have made significant strides over the past decade in diagnostic biotechnology fields. STD might give Monotech knowledge and access to technology developed in Japan that it might not find out about, or have an inside track to license, otherwise. In this way Monotech might use STD to monitor and acquire Japanese technology, help it commercialize, and exploit the capabilities of the venture in the global market. It would also improve Monotech's distribution and technology presence in Japan.

Monotech has been discussing issues broader than current product and market concerns with STD and the Japanese parents over the past several years. But there are obstacles to Monotech becoming more involved. For example, an equity stake for Monotech in the venture would be desirable and perhaps necessary to generate the synergies discussed above. Competition between the Japanese parents would complicate the negotiations toward a fundamental restructuring of this kind.

It may or may not be possible for Monotech to change the structure of the relationship and upgrade its involvement. To Joseph Atkins, president of Monotech's Diagnostics Division, the experience of American Medical Equipment (AME) in Japan is relevant to Monotech's current situation. AME had a joint venture with Oda Denki in computed tomography (CT) scanners, which AME played a major role in pioneering. The Japanese market grew very rapidly, and the joint venture was very profitable because of AME's technological lead. But over time Japanese competitors caught up to AME by developing products with features (basically smaller size) that were tailored to the Japanese market. For its part, Oda wanted to integrate into manufacturing and demonstrated an ability to reverse engineer all of the scanner's hardware. What Oda was not able to duplicate was AME's software for the scanner. If Oda had had its way at that point, it might have chosen to license the software and take over manufacturing itself, but AME was able to leverage its technology in order to increase its involvement in the joint venture. It was a 51-49 venture in favor of AME, and over the past 8 years AME's ownership has moved to 75 percent, market share in Japan has solidified, and the partners have closer technological links. The venture has grown its own technology, and AME now exports mid-sized machines from Japan as the global CT scanner market has stratified.

Monotech may not have as much leverage now as AME had in the early 1980s. But the company has some new products in the pipeline for diagnosing lung, breast, and bladder cancers. With the growing emphasis on preventative health care, the Diagnostics Division sees its products as well positioned for long-term growth. And just as the Japanese market is critical for the long-term growth of the Diagnostics Division and Monotech as a company, Showa sees its relationship with Monotech as an important component in its long-term strategy to build a larger presence in health care.

The Monotech-STD linkage may be in a period of transition. The case illustrates how in some applications of biotechnology cooperation and competition can be managed, though the process is often complicated. Another theme is how an emerging U.S. biotechnology company can get the most mileage out of its fundamental strategic asset_technology. Through awareness of changes in the global market and maintenance of a technical edge, American biotechnology companies may be able to define and accomplish changes in the structure of their technology linkages to Japan in order to ensure maximum leverage and increasing benefits over time.

CASE III: KIRIN-AMGEN

Kirin-Amgen, Inc., the joint venture established in 1984 to develop and market erythropoietin (EPO) and, later, granulocyte colony-stimulating factor (G-CSF), is perhaps the best known and most successful U.S.-Japan biotechnology linkage. Kirin-Amgen persevered through over 5 years of product development, clinical trial management, and building manufacturing and marketing capability before EPO was approved for sale in the United States in June 1989. The two products have met with resounding success. Because of the significant and obvious benefits that have accrued to both sides, some have pointed to Kirin-Amgen as a model for linkages between emerging U.S. biotechnology firms and large Japanese corporate partners. However, with two blockbuster products on the market, the venture may be facing a dry spell in its development pipeline. At this point it is uncertain whether Kirin-Amgen will continue to play as prominent a role in the strategies of the partners.

The Partners

Kirin undertook a major diversification program in the early 1980s (see Case I). Pharmaceuticals have received particular emphasis in the program. The long-term plan put forward in 1981 gave several reasons for entering the pharmaceuticals business, including the industry's knowledge-intensiveness. The technological basis of pharmaceuticals had some similarity to those of existing businesses, and the "new biotechnology" gave Kirin an opportunity for market entry in a technical field where the more established Japanese pharmaceutical companies were just as inexperienced as it was. Kirin hoped to leverage its accumulated fermentation, biochemical, and engineering expertise to build a technical critical mass for biotechnology. The company also hoped to establish an international information network that would speed the identification of promising technologies. A corporate Technology Information Division was established in 1986, and about 70 percent of its personnel are devoted to pharmaceutical activities.

Amgen, Inc., based in Thousand Oaks, California, was founded in 1980. George Rathmann, who was previously vice president for R&D at Abbott Laboratories, was hired as the company's first CEO and served in that capacity until 1988, when current CEO Gordon Binder took over. Because of the success of its first two products, Epogen and Neupogen (the brand names given to EPO and G-CSF), Amgen's quarterly revenues passed Genentech's during 1991, and the firm is now the sales leader among dedicated biotechnology companies. It is anticipated that Amgen will be the first biotechnology company in the Fortune 500. Amgen had sales of $361 million during its 1991 fiscal year (which ended March 31, 1991) and it employs 1,179. The company spent almost $85 million on R&D in fiscal year 1991.

Amgen's success is based on two proteins, EPO and G-CSF. EPO is a protein that stimulates the production of red blood cells and replaces blood transfusions in the treatment of kidney dialysis patients and patients with other indications. EPO is produced naturally in the kidneys. It received regulatory approval in Europe in 1988 and U.S. FDA approval in 1989 and was approved in Japan in early 1990.

G-CSF is one of a class of colony-stimulating factors that "control the differentiation, growth and activity of white blood cells."[80] G-CSF stimulates the production of neutrophils, white blood cells that fight infections. It was approved by the FDA in early 1991 for use by patients undergoing chemotherapy. It received approval in Europe at about the same time, and was waiting for approval in Japan at the time of this printing. Amgen is taking the drug through clinical trials for other indications, such as burn cases and pneumonia.

Origins of the Linkage and Negotiations

In February 1984 George Rathmann received a telephone call from a Kirin representative who wanted to know why Rathmann was not answering

---

[80] Ann M. Thayer, "Biopharmaceuticals Overcoming Market Hurdles," *Chemical and Engineering News*, February 25, 1991, p. 38.

the numerous telex messages that Kirin had sent to him. It turned out that Kirin was using the wrong telex number, and that the Japanese firm wanted to set up an appointment the following week to discuss licensing rights to EPO. Amgen's Fu-Kuen Lin succeeded in cloning EPO in October 1983, and Kirin scientists had read an announcement about it. Kirin decided that it wanted to be in biotechnology, thought that EPO was an interesting product, and decided to try to acquire the rights to it.

Not many other large companies were interested in EPO at the time. There were several reasons for the early lack of interest, which seems surprising on the surface because of the drug's subsequent success. To begin with, even after the protein was cloned, there was doubt about whether EPO would be efficacious with no side effects, which turned out to be the case. Further, the economic feasibility of producing an effective dosage at a price that would represent a savings over the blood transfusions that EPO treatment would replace was another area of considerable uncertainty. Also, EPO is an injectable, and few pharmaceutical products that are limited to that delivery system have achieved prominent success. Finally, EPO represents a significant change in therapeutic approach_a significant advance as it turns out_but it was difficult to estimate the overall size of the potential market a priori because there were no competing drugs on the market.

Rathmann met with Kirin's representatives in March 1984. The basic concept for the partnership was arrived at then. Amgen first proposed, without citing specific figures, an exclusive Japanese license for Kirin in exchange for a front-end payment and a significant royalty. Kirin insisted on more than marketing rights in Japan. The negotiations might have ended there, but Amgen responded by proposing that EPO be developed and marketed in a joint venture. The main advantage of a joint venture was that risk and return sharing would be self-compensating for unknowns on both the downside and the upside. In contrast, a license and its accompanying royalty rate assume something about market size and profit margins. In a 50-50 joint venture, the partners would share equally in the costs if it proved to be more expensive than expected to take EPO through clinical trials and would likewise share the benefits if the sales were higher than anticipated.

Kirin was amenable to the basic concept, which included some adjustments to a basic 50-50 structure. At the outset Kirin put up $12 million and Amgen put up $4 million because Amgen was contributing the fundamental technology of manufacturing EPO. It was initially anticipated that most of this start-up capital would be spent getting the drug through clinical trials in the United States; the Japanese approval process was expected to drag on for a longer period but to be cheaper. The venture would be managed by a board composed of three representatives from each company, with the president/CEO post held by Amgen and the chairman slot controlled by Kirin.

Protection for Kirin was built into the joint venture as well. The main point in question at the outset was whether the manufacturing technique could be refined to the point where making EPO from host cells would be economically feasible. It was agreed that Amgen would spend its $4 million contribution on bringing the production method to the point of economic feasibility within 18 months. If this could not be accomplished within the specified resource constraints, Kirin would have the option of taking its $12 million and walking away. The level of "feasibility" that was arrived at was about 50 times the level of efficiency that was being achieved at the time of the preliminary agreement, in the spring of 1984. Amgen passed this milestone within several months, the technology transfer of the methods for producing host cells and for using the cells to manufacture EPO was accomplished in late 1984, and Kirin committed its initial $12 million. Rathmann served as president/CEO of Kirin-Amgen until 1988, when Gordon Binder replaced him. Yasushi Yamamoto represents Kirin's board of directors as chairman of the joint venture, replacing Dr. Kubo.

Structure and Evolution

Through 1988 the management of the venture went smoothly. The board made no decisions that were not unanimous, which might be expected in a 50-50 joint venture. Though the chairman leads the board of directors, which has formal supreme authority, the president/CEO is responsible for managing the venture.

The joint venture's main function has been to manage the technical exchange and product development collaboration between the partners. Since the actual R&D is done by the partners themselves in exchange for a fee charged to the joint venture, in practice Kirin-Amgen assumed a largely planning and monitoring role.

The partners have different marketing arrangements for EPO. Sankyo is marketing the drug for Kirin in Japan for all indications, and the Ortho Pharmaceuticals subsidiary of Johnson & Johnson is seeking approval for nondialysis indications in the United States. Amgen was granted a 7-year monopoly in the dialysis market under the Orphan Drug Act. At one time Kirin-Amgen owned all the rights to EPO, but the rights were transferred back to Amgen in the U.S. market and to Kirin in the Japanese market. Amgen has waged a long and complicated legal battle with Genetics Institute and its licensee, Japan's Chugai Pharmaceutical, over patents for EPO. Kirin has Japanese rights, Amgen has rights in the United States, and the venture owns rights in other markets. Kirin and Sankyo face competition in the Japanese market, where Chugai accounted for over half the sales of EPO as of mid-1991. Originally, a 5 percent royalty on all sales of EPO by the partners and their licensees was to go to the joint venture. However, it was later decided that only dialysis sales in the United States and Japan would be royalty bearing to Kirin-Amgen and that EPO marketed for other indications would not be royalty bearing. Kirin manufactures EPO in Japan, while Amgen manufactures it in the United States for the American market.

In late 1984 the partners began discussions about the inclusion of other molecules in joint venture product development. Kirin was investigating a protein called thrombopoietin, and Amgen was exploring G-CSF with Sloan-Kettering. Kirin proposed developing both of them within Kirin-Amgen. Amgen was agreeable but wanted to pursue the independent programs for a while longer. Amgen spent more on its program and, ultimately, G-CSF was successful while work on thrombopoietin has thus far produced less promising results.

In the summer of 1985, Amgen agreed to put G-CSF into the joint venture. The partners agreed on a 50-50 split outside the United States and Japan. This arrangement was changed several years later. Amgen proposed that either the two companies agree to enter the European market together or that Amgen should be allowed to buy back the European rights. The two sides agreed on the latter course, and Amgen bought back the European rights in 1986. The two companies established marketing rights in other areas, with Kirin getting Taiwan and Korea and Amgen getting Australia and Canada. Amgen comarkets Neupogen with Hoffmann La Roche in several of its territories outside the United States.

As with EPO, Kirin insisted on manufacturing the G-CSF that it would market. There are manufacturing facilities in the United States and Japan for both products that were financed and owned by the partners individually. Development and clinical trials were financed by the joint venture. During 1991 Kirin-Amgen paid Amgen $14.6 million for contract research, and Amgen paid the venture $17.1 million in royalties on sales of Epogen and Neupogen.

Technology Transfer

All the technology transfer was accomplished at the science and engineering level between the two partners, mainly through visits of Kirin research personnel to Amgen. At times when contact was most extensive, four Kirin researchers at any one time were posted at Amgen, with some staying as long as 3 years. Kirin saw this as a particularly beneficial component of the relationship. Much of the technology exchanged related to the proper way to treat host cells in order to maximize the production of EPO. There were also visits by Amgen researchers to Kirin for periods of up to a month. Techniques for the purification of EPO were transferred to Kirin as they were improved upon. Amgen also provided Kirin with basic materials from its cell bank, such as cloned cells, for use in research and manufacturing.

After the initial hurdle of commercially feasible efficiency was cleared, the focus of product development moved to clinical trials in various countries. Technical personnel and information also were exchanged during all phases of the design and construction of the manufacturing facilities in the United States and Japan.

Press reports at the time Kirin-Amgen was launched speculated that one reason Amgen chose to team up with Kirin was the latter's fermentation technology and potential to develop bioprocessing skills. In fact, the differences between fermenting beer and bioprocessing cloned proteins are so great that no technical synergy could realistically be expected. Amgen's main reason for linking with Kirin was the Japanese firm's interest in EPO. Though Kirin actively contributed to the process of developing EPO, the bulk of the know-how that came to be used for treating the host cells was developed by Amgen.

Still, important technological contributions from Kirin did materialize. To begin with, the collaboration during clinical trials brought substantial benefits. Kirin handled all the clinical trials in Japan, for both EPO and G-CSF, including animal trials. These are not required for the FDA, but they were used in the United States and other parts of the world to bolster the case with regulators. Perhaps of even greater benefit was the establishment of a "world view" during the clinical trial process, which provides something of a three-dimensional perspective on the clinical importance of the drug. The fact that leading clinicians all over the world were working on the trials simultaneously provided a tremendous check and probably saved a significant amount of time. With different teams working simultaneously and exchanging information, clinical interpretation and documentation were validated to an extent that would be impossible in one place.

It was also particularly important to the process of regulatory approval for the manufacturing techniques to be translatable between the partners. This is because EPO has five isoforms, and it is critical that the partners be able to show regulators that the proportions are standardized for the purpose of evaluating efficacy and side effects. Particularly in a case where companies are working to improve the efficiency of a manufacturing process, knowledge may be accumulated and not written down. That the joint venture was able to prevent such "droppage" is perhaps due to the effectiveness of Amgen's technology transfer to Kirin.

Kirin made one unexpected technical contribution when the partners were designing their individual manufacturing facilities for EPO. Kirin and Amgen consulted closely during this stage. Amgen had wanted to use a manufacturing process in which "roller bottles" wash nutrients over the genetically engineered host cells that produce EPO. Kirin believed that the roller bottle process could be automated and with the help of one of its suppliers was able to develop a machine that handles the roller bottles. An

Amgen employee visited Japan for about a month to learn the specifics of running the machine. The automated process is used at Kirin, Amgen, and Johnson & Johnson in making EPO. The manufacture of G-CSF uses a different process.

Impacts and Factors Contributing to Success

The commitment of the top managers of both companies throughout has clearly been critical to the success of the venture, particularly through the long period during which resources had to be expended in the absence of a revenue flow. In contrast to the usual view that Japanese management has a very long-term outlook and that it is the American partner to a joint venture whose commitment is likely to waver, it may have been more difficult at times for Kirin to maintain the degree of commitment that it did. The core business of the company is still clearly brewing and selling beer. Yet Kirin did manage to maintain an unwavering focus on the venture in terms of resources and attention.

A number of significant benefits have accrued to Kirin as a result of the venture. First and foremost, Kirin was able to break into the ethical drug market in Japan with two hit products. The Japanese partner had a long wait, but it is now enjoying high returns on its investment. Kirin was also able to achieve its stated goal of technology leveraging. Finally, the company was able to use the experience gained by its technical personnel through Kirin-Amgen to establish a basic research facility in the United States. Kirin took this step in 1988, with the establishment of the LaJolla Allergy and Immunology Institute.

After Kirin-Amgen's initial $16 million in capital was used up, it took over $80 million more to take EPO through to FDA approval. Those costs were split equally between the partners. Most of the interaction has occurred at the scientific level, and Amgen credits Kirin for the quality of its technical leadership. For Amgen, maintaining its focus on the venture was a straightforward proposition. EPO was the company's flagship product and main hope to become an independent pharmaceuticals company. Kirin-Amgen was the vehicle chosen for its commercialization, and commitment to the venture had to be maintained to bring the product to market.

Recent Developments and Prospects for the Future

Though there are no new products in the Kirin-Amgen pipeline, there is ongoing technical interaction. This mostly involves clinical studies on new indications.

The Kirin-Amgen joint venture continues to collect royalties from sales of the two products and passes these on to the partners and will continue to do so for many years. At the same time, the goals and priorities of the companies are evolving. Over the past several years Kirin has given more attention to the Japanese beer market, as domestic competitors have mounted a serious challenge. For its part, Amgen is exploring several partnerships with smaller U.S. biotechnology companies, in addition to the work it is doing internally, as a means of expanding its product line.

Clearly, the joint venture has addressed some of the complementary needs of the two partners and balanced the asymmetrical assets that drove the Kirin-Amgen partnership at the outset. Both sides now have a wider range of options. Benefits will continue to flow to the partners, but the future importance of Kirin-Amgen in their strategies is unclear.

CASE IV: HITACHI CHEMICAL RESEARCH_ UNIVERSITY OF CALIFORNIA, IRVINE

The agreement between Hitachi Chemical Research (HCR), a subsidiary of Hitachi Chemical Company, Ltd., and the University of California, Irvine (UCI), to occupy the same building on the UCI campus is part of a clear trend toward increasing interaction between Japanese corporations and American academic research institutions. This trend is apparent in biotechnology and in other fields, such as computer sciences and electronics.

However, this particular linkage represents something of a departure from traditional relationships between U.S. universities and corporations. The foundation of the interaction is an exchange of leases: HCR built the facility on university-owned land and has established a basic research lab for its proprietary programs on the top two floors; in return, UCI's Department of Biological Chemistry occupies rent-free lab and office space on the first floor of the building and will take over the entire building in 2030. The corporate and university researchers share a reading room. It is expected that their physical proximity will facilitate formal research interaction, but this will be managed on a project-by-project basis on the same terms that normally govern UCI collaboration with industry. HCR has no formal funding commitment to UCI, and UCI made no commitment to HCR.

Research at the lab began in the spring of 1990, so it is too early to assess many of the impacts on Hitachi Chemical and UCI. What can be said now is that the partners are working hard to ensure that the relationship benefits both sides, and they hope that the structure of their agreement and the process for reaching it can serve as something of a model for new forms of university-industry research interaction.

The Partners

Hitachi Chemical Research is a wholly-owned U.S. subsidiary of Hitachi Chemical Company, Ltd. Hitachi Chemical has traditionally focused on developing and manufacturing synthetic resins for applications in electronics and also produces molded parts for automobiles and housing equipment. The company is emphasizing new ceramics and is seeking to diversify into the pharmaceuticals business. Hitachi, Ltd., owns over 50 percent of Hitachi Chemical. In the fiscal year ending March 1990, Hitachi Chemical registered sales of over 466 billion yen ($3.5 billion at 135 yen per dollar), earned 19 billion yen in operating profit ($145 million), and spent over 12 billion yen ($92 million, or 2.6 percent of sales) on R&D.

The University of California at Irvine is one of nine campuses in the UC system. The university manages three U.S. Department of Energy laboratories. University-wide, the system receives about $4 billion in extramural research funding per year and has 10,000 faculty members. At the Irvine campus, industrial sponsors provide between 5 and 7 percent of the research funding in a given year. Most research performed at UCI, biochemical research in particular, is funded by the federal government through NSF and NIH. Federal funding to the nine campuses accounts for over 10 percent of the federal budget for academic research. Biochemical research connected with the medical school has increased substantially over the past decade. The Department of Biological Chemistry has 10 faculty members and receives extramural research funding of over $2 million per year.

The agreement with Hitachi Chemical is one of a number of relationships that UCI has entered into over the past decade with the view of using its land resources to meet priority academic needs. UCI has lease-swap arrangements similar to the HCR lab with several nonprofit organizations such as Beckman Laser and the American Heart Association. In addition, the university previously leased land to another private pharmaceuticals concern, the Nelson Research and Development Company, allowing it to build an R&D facility in exchange for use of space in the new building by the Department of Psychiatry. That relationship was established in the mid-1980s but was dissolved after about 5 years when Nelson was acquired by Ethyl Corporation and its research operations were moved to Richmond, Virginia. UCI bought out the remainder of the Nelson lease and now uses the entire building.

UCI uses its land to meet other needs besides research space, one example being the construction of housing for sale to faculty members with shared appreciation, thus making it possible to sell the housing at a lower cost than market value for fee-simple housing. This is an important recruiting tool given the high cost of housing in Southern California.

The Origins of the Linkage

In 1984 UCI was recruiting Professor Masayasu Nomura, a prominent Japanese biochemist, and was faced with the problem of providing enough research and office space for him and for the projected future growth of the biochemistry research faculty. In the course of his recruitment, contact was made with Hitachi Chemical, which agreed to support an endowed chair. Subsequently, this led to an expansion of the relationship, which culminated in the building. Research at the building Nelson constructed on campus was just getting started at this time, so UCI had an existing prototype for an arrangement that would match complementary needs and resources. Leases had already been written that could be adapted to the particular case, and the campus also had accumulated experience in negotiating agreements for shared facilities, which was useful in dealing with Hitachi Chemical.

Negotiation Process and Issues

Hitachi Chemical was receptive to the basic concept of a swap of leases and the construction of a shared research facility at the outset, and it set up the Hitachi Chemical Research subsidiary to negotiate an agreement and manage the lab. The basic agreement being contemplated was, indeed still is, quite unusual in the context of university-industry research. Some issues would likely arise regardless of the nationality of the company, while others were specific to HCR's parent being Japanese. In the negotiation process, the campus coordinated interaction with HCR and the UC president's office and legal counsel. This contributed to the smooth management of the negotiation and implementation process and a focused effort to ensure maximum benefits for the campus within the framework of university policy.

UCI was perhaps most concerned about potential political repercussions. In 1987 concerns were being raised in Congress and elsewhere about the potential adverse impacts on U.S. competitiveness of Japanese and other foreign corporate involvement in U.S. academic research. Public universities are supported by their respective state governments, and federal funding is critical to the research enterprise. For a number of large public research universities, federal support is comparable to or exceeds state support. Political concerns have been raised about relationships between foreign companies and research institutions supported with public funding.

At the federal level, some members of Congress are concerned that the open research policies of U.S. universities combined with the willingness of foreign companies to invest in U.S. academic research can, in effect, translate into subsidizing foreign industry in an increasingly competitive global economic environment. Some focus on relationships with Japanese companies, asserting that comparable benefits are unavailable to U.S. companies operating in Japan and that the ability of Japanese companies to access U.S. academic research allows them to "free ride" on U.S. basic research.

Political concerns also arise at the state level. California, like other state governments, encourages business development in biotechnology and in other growth industries. Some programs seek to leverage the research capability present in the public university research system. Some would argue against the UCI-HCR relationship on the grounds that research interaction between state universities and California firms should take priority over interaction between universities and foreign companies. On the other hand, in an example of the mixed signals that academic institutions sometimes receive when it comes to Japan, California is trying to position itself as a key participant in the emerging Pacific Rim economy. Restrictions on relationships such as the one between UCI and HCR could be difficult to reconcile with such a stance.

Although UCI officials went out of their way to ensure that Hitachi Chemical was given no consideration that would not be extended to domestic or local companies, they were concerned that it might be perceived that they had done so. The university realized that very few, if any, U.S. companies would be able or willing to make the large long-term commitment necessary to build the UCI-HCR facility and operate it. The university decided that the best way it could allay concerns would be to ensure that its relationship with HCR was governed by normal university procedures and that HCR did not receive special treatment regarding intellectual property rights or other areas. While it is generally known that UCI can entertain land for space arrangements with domestic industrial sponsors, a general solicitation was not judged necessary in view of the fact that UCI was happy to entertain any proposal from any company and that it was not negotiating a unique lease.

Building the Facility

The UC regents approved the project in March 1988. Ground was broken on the $12 million, 40,000-square-foot facility in January 1989, with UCI and Hitachi Chemical taking occupancy in the summer of 1990. Over the 18 months from the time ground was broken to when the building was occupied, HCR and UCI interacted closely on design, construction, and outfitting the building. The parties also defined the procedures and responsibilities for interaction.

Before the Hitachi building was constructed, the Department of Biological Chemistry's research facilities were scattered in three locations. One of UCI's goals was to bring the department together. This influenced the layout and which faculty members moved into the new building. The first floor of the Hitachi building houses two faculty members with large research groups and one with a smaller group. This floor is connected by a corridor to the rest of the biochemistry research facilities.

Because it was known beforehand who would be moving, it was possible to design the space according to the needs of the users. Since the university had no budget for design and construction (there was a small administrative budget) and it would be responsible for any changes that it initiated in the design, it was desirable for UCI to clarify the design parameters at an early stage. This was largely accomplished; the university initiated very few changes in design.

The university's goal was for a "turnkey" facility. Hitachi agreed to provide all equipment defined as "nonremovable," including cold rooms, the ionized water system, compressors, and an emergency generator. For one piece of equipment, Hitachi had planned to use its Japanese supplier, but UCI researchers explained that it needed equipment with higher specifications, and Hitachi switched to UCI's U.S. supplier. HCR pays two-thirds of the costs of maintaining these common facilities, while UCI pays one-third. Decision making on building and facilities matters is accomplished by a joint committee.

The reading room on the second floor is shared. Hitachi also has a lecture room, which it allows the university to use on request. There is no formal agreement, but UCI sometimes holds classes in this room.

One issue that arose during construction had nothing to do with research interaction or the fact that HCR is a subsidiary of a Japanese corporation. This was the question of insuring the lab, where toxic and hazardous materials might be used for research. HCR eventually agreed that it would self-insure through a $5 million escrow account. The issues of ensuring compliance with state regulations regarding hazardous materials, animal research, and other areas must be dealt with whenever a chemical or biotechnology research lab is constructed. Complications may arise in determining responsibilities in a novel university-industry partnership such as this one.

Guidelines for Interaction

UCI's Office of University/Industry Research and Technology, in the course of implementing the agreement, placed a top priority on making sure that university policies in a number of areas were clear and mutually understood. University policies that govern research interaction are covered in a number of documents and come to several hundred pages in all. Federal and state regulations have an impact as well. The office summarized these policies in a document called "Guidelines for Research Interaction," which was agreed to by both UCI and HCR and was to have been distributed to all occupants of the building in 1991. The document spells out the roles of a number of university offices and academic departments as well as the appropriate procedures for initiating, implementing, and managing various forms of research interaction between UCI and HCR. These include sponsored research, consulting agreements with faculty, transfer of research materials, use by one party of the other's facilities, and services for fee. The guidelines encourage the use of explicit written agreements by the parties to govern interaction. It might be useful to examine how the guidelines treat aspects of interaction related to intellectual property rights, export controls, and conflicts of interest.

Hitachi's lab and the UCI researchers on the first floor are conducting research on biochemistry with a view toward possible human health care applications. In biotechnology, basic science and applied research are often quite close. Furthermore, intellectual property rights are usually a more important consideration for companies in pharmaceuticals than in other industries. Therefore, the topic receives a great deal of attention in structuring research relationships between companies and universities in this field. The university has an interest in making sure that it owns the rights to commercially valuable technology developed in its labs while simultaneously maintaining academic freedom.

The standard University of California approach to intellectual property rights is university ownership of all research results produced using university resources. At a state university, research is ultimately aimed at benefiting the public, with the generation of income as a secondary priority, so intellectual property rights (IPR) and licensing policies are not as flexible as they are at some private institutions. UC's policy states that all university employees as well as "all noncompensated persons who use university resources," including facilities and equipment, are required to sign the university's patent agreement. A key element in the UCI-HCR linkage is that the Hitachi portion of the building and the land are not considered "university resources," so that Hitachi owns all intellectual property developed on its floors. The floor occupied by UCI faculty is considered university property. The policy also states that, in the case of "joint inventions of at least one UCI inventor and at least one HCR inventor," UCI and HCR will each own an equal interest in the invention.

HCR or any other research sponsor that pays all direct and indirect research costs may be granted the first right to negotiate an exclusive license. In cases where HCR supports research along with other sponsors, the company may be granted the right to negotiate a nonexclusive license. A sublicense to the parent company, Hitachi Chemical, can be considered in this situation as well. HCR, in addition to technology developed in projects that it sponsors itself, may also be allowed to license technology that arises from other research at UCI.

The university may grant a short delay to HCR for filing patent applications before researchers publish the results of research that HCR sponsors. Unlike the United States, most countries use a first-to-file patent system in which the first applicant is granted the patent, assuming other requirements for patentability are met, so it is important to file before results are published. UCI may give this delay by agreement but if necessary would forgo foreign rights to technology rather than compromise academic freedom by delaying publication too long.

In all licensing negotiations for university-owned intellectual property, the prospective licensee must submit a business plan for commercializing the technology, which UC then evaluates. It would be unusual for a company to sponsor research that it is clearly incapable of commercializing, so this evaluation normally does not raise obstacles.

A second issue covered in the guidelines is compliance with federal regulations regarding export controls, since specific licenses from the government may be necessary prior to exporting a technology or filing foreign patent applications. Since UCI has contracted with HCR, a U.S. company incorporated in California, it is specified in the ground lease that responsibility for compliance with export control regulations lies with HCR. In addition, UCI's policy is that only fundamental research is conducted on its floor, to ensure compliance with federal policy allowing fundamental research to remain open and unclassified.

A final issue is faculty consulting. Most faculty members and nearly all of the senior professors at the UCI medical school have some relationship with corporations through research sponsorship or consulting. Universities often encourage this activity because it gives faculty insight into the types of technical problems faced by industry.

From the university standpoint, the main concern is avoiding potentially harmful conflicts of interest. These would arise when a faculty member has a major financial interest in a corporate sponsor of his or her university research. When HCR proposes sponsored research at UCI, faculty members receiving the grant must disclose any financial interest in HCR or its parent to the UCI Conflict of Interest Oversight Committee. The university needs to consider a number of factors, including whether HCR and UCI's activities are being kept separate, whether the faculty member participated in the decision to make the award, and ensuring the openness of the university research environment. The committee uses this information to decide whether the award should be taken.

Impacts and Prospects for the Future

Jack Jacobs, science director of the HCR R&D lab, previously worked at Merck and has a great deal of experience in conducting research in collaboration with universities. He holds a UCI adjunct faculty appointment in the Department of Biological Chemistry. There are over 20 HCR staff members on the top two floors of the building. HCR hired three researchers from UCI_by mutual agreement with the university_from a program that experienced a reduction in its funding. None of the researchers were tenured faculty. For HCR, "raiding" UCI of its top researchers would not contribute toward good long-term relations with the university. When recruiting researchers, Jacobs can raise the possibility that working at HCR may facilitate an adjunct faculty appointment subject to UCI's requirements and approval.

The three UCI research groups on the first floor do basic biochemical research representative of what is being done in many research universities. One group is working on mapping human chromosomes, which may have implications for pinpointing and curing genetic disorders. The other groups are working in more basic areas_RNA processing in yeast and the organization and biosynthesis of ribosomes. The work is presently funded by a variety of NIH, NSF, and private foundation grants.

Interaction between the UCI faculty members housed in the building and HCR personnel has not been extensive thus far. Indeed, the first research contract, invention, and licensing agreement between UCI and HCR resulted from a project that Hitachi sponsored for UCI's Department of Pharmacology, which is not housed in the HCR building, and the University of Oregon. UCI and HCR have also established a standard request form letter that UCI researchers can complete in order to use advanced equipment located on the HCR floors.

For the university, the positive impacts are fairly straightforward. A number have already been realized. The university has the use of high-quality space more quickly and at a lower cost than if a lab had been built through state support. Quality facilities of this type allow faculty members to be more productive and improve the quality of graduate education. Also, to the extent that HCR gives assistance or research positions to graduate students, this will allow the graduate school to train more students.

Direct sponsorship of research by HCR that arises from the physical proximity is an expected benefit, though the extent is not yet clear. There are also spin-off benefits from sponsored research. When HCR agrees to license technology, it pays for the patent applications and the issue fee for the patent as well as the royalty on sales of the commercial product. The university has not encountered negative impacts from the agreement and does not anticipate any, though it should be pointed out that research at the facility had been going on for about a year and a half as of this writing.

There are no concrete plans to make closer interaction between UCI and Hitachi's Japanese biotechnology lab or other Japanese research institutions a formal part of the relationship. Professor Nomura has strong ties to the research community in his native country, and it is expected that personal relationships may lead to closer UCI-Japan interactions over time.

UCI sees this relationship as an innovation in the structuring of university-industry research interaction that benefits itself, HCR, and U.S. biotechnology as a whole. The campus hopes that new relationships with industry will further cement its position as a leading-edge research institution. New modes of cooperation with industry that are likely to arise in coming years will require universities to consider issues beyond licensing, and UCI is pleased with the results of its coordinated approach.

For HCR and its Japanese parent, the impacts are perhaps less clear at this point. From the point of view of the company, this linkage represents a leap into uncharted territory in a number of respects. For example, biotechnology is a relatively new technological field for Hitachi Chemical and pharmaceuticals are a new business. In addition, the decision to launch into biopharmaceuticals by building a basic research capability represents a departure from the approach taken by Kirin and other large Japanese companies, in which a gradual shift of research focus was accompanied by linkages with U.S. biotechnology firms to obtain product rights and technology closer to the commercialization stage. Finally, focusing a basic research thrust on a laboratory in the United States and a novel relationship with a U.S. university will present Hitachi Chemical and the HCR subsidiary with an additional layer of organizational and business challenges.

It might be expected that a combination of "trial and error" and "learning and listening" will prevail at HCR for the time being. At this point, the facility is mostly staffed by U.S. researchers with academic and corporate backgrounds. The lab does have the potential to play a key role in building the parent company's biotechnology capability by serving as a training ground for Japanese researchers, who could familiarize themselves with methodology and developments in U.S. biotechnology through short- and long-term visits. In addition, the fact that a technology with commercial potential has already been developed through this relationship points to the possibility of a substantial payoff in the long run from an investment that very few U.S. companies in the pharmaceuticals industry would be willing to make.

APPENDIX B

Examples of Linkages Between Japanese Companies and U.S. Academic Research Institutions

Year	Japanese Partner	U.S. Partner	Type of Linkage	Technologies	Product	Comments
1982	Green Cross	University of California	Collaborative research	Hybridoma	Monoclonal antibodies (MABs) therapeutics (RX) for cancer	Agreement for development of MABs for cancer
	Toyo Jozo	Johns Hopkins University	Licensing agreement	Recombinant DNA (rDNA)	Interferon (IF)	Toyo has licensed IF technology from Johns Hopkins
1983	Asahi Chemical Industry	City of Hope Medical Center	Collaborative research	rDNA	Interferon-G (IF-G)	Asahi will build a large production tank for its IF-G developed through collaboration with City of Hope
	Chugai Pharmaceutical	University of South Carolina	Collaborative research	Hybridoma	MABs	USC gets $500K over 3 years for MABs and cancer diagnostics (DX) development; lymphokines
	Sumitomo Chemical	U.S. Cancer Research Center	Collaborative research	rDNA	Macrophage activating factor (MAF)	Sumitomo will get use of MAF developed in the United States

1984	Suntory	New York State University	Collaborative research	rDNA	Plasmids	NYSU will supply Suntory with plasmids for rDNA studies
	Suntory	Rockefeller University	Collaborative research	rDNA	RX for dementia	Joint development of an RX for senile dementia
	Toyo Jozo	New York State University	Collaborative research	rDNA	Oncogene	Two-year joint research agreement to study carcinogenic gene
1985	Ajinomoto	MIT	Collaborative research		Cell biology and immunology	Ajinomoto provides $750K annually for 5 years
	Mitsui Toatsu Chemicals	Beckman Research Institute	Collaborative research	rDNA	Tissue plasminogen activator (TPA)	Mitsui Toatsu is working with Beckman Research to produce TPA in animal host cells
1986	Otsuka Pharmaceutical	Fred Hutchinson Cancer Research Center	Licensing agreement	MABs for cancer DX		

APPENDIX B *Continued*

Year	Japanese Partner	U.S. Partner	Type of Linkage	Technologies	Product	Comments
1986_ *cont'd*	Takeda	Harvard University	Collaborative research		Renewal factors	Takeda funds $3 million over 3 years to Harvard University Childrens' Hospital to study blood vessel renewal factors and inhibitors affecting cancer metastasis and bone formation
1987	Konishiroku Photo	Stanford University	Collaborative research		Tumor marker	Joint venture that has discovered tumor marker believed to be common to most cancerous cells; isolated and purified a glycosyl-transferase
	Lyphomed	Michigan State University	Collaborative research	rDNA	Antifungal antibiotic	Further development of an antifungal antibiotic

Year	Company	Institution	Type	Product	Description
	Otsuka Pharmaceutical	University of Maryland	Collaborative research	Interleukin-1 (IL-1)	Joint project for protein engineering of IL-1; will be used as RX in various infectious diseases and to reduce side effects in radiation therapy
1988	Hitachi Chemical	University of California, Irvine	New research facility		Hitachi built lab on UCI campus; in return, UCI receives use of one floor of lab space
	Kirin Brewery	University of California, Santa Barbara	Collaborative research	Megakaryocyte colony stimulating factor (MEG-CSF)	Collaboration to develop MEG-CSF for treatment of thrombocytopenia
	Nagoya Sogo Bank	Columbia University	Gift	rDNA	$50K to life sciences fund
1989	Asahi Chemical Industry	SIBIA	Collaborative research	rDNA seeds	Asahi signed agreement with Salk Institute Biotechnology Industrial Associates to develop new fruits and vegetables via rDNA

APPENDIX B Continued

Year	Japanese Partner	U.S. Partner	Type of Linkage	Technologies	Product	Comments
1989_ cont'd	Daiichi Pharmaceutical	Vanderbilt University	Endowed chair			$1.2 million; also provides for exchange of research staff
	Mitsubishi Chemical Industries	McLean Hospital	Collaborative research		RX for Alzheimer's disease	Mitsubishi will fund 200 million yen over 3 years to jointly develop RX for Alzheimer's disease
	Shiseido	Massachusetts General Hospital	New research facility		Dermatology research	Shiseido will provide $85 million to establish world's first comprehensive dermatology center at MGH called the Harvard Cutaneous Biology Research Center
	Sumitomo Chemical	SIBIA	Collaborative research		Disease/pest-resistant plants	Sumitomo provides $900K in a 2-year cooperative plant

1990	Fujisawa	University of Pittsburgh	Collaborative research	Immuno-suppressants	Joint development and clinical testing of immuno-suppressants
	Japan Research and Development Corporation	Michigan State University	Collaborative research	Environmental biotechnology	Five years and $15 million toward evolution of microbes for environmental biotechnology
1991	Daiichi Pharmaceutical	National Cancer Institute	Supply drug for testing		Daiichi will supply SP-PG, a treatment for Kaposi's sarcoma to NCI for testing and clinical trails
	Kanebo	University of Pittsburgh	Collaborative research	Anticancer agent	Five-year study of anticancer and immune system technologies
	Takeda	Harvard University	Collaborative research	Anticancer drug	Fumagillin set to enter U.S. clinical trails

(continued from previous row) research agreement with SIBIA; SIBIA will develop disease/pest-resistant plants

APPENDIX B Continued

Year	Japanese Partner	U.S. Partner	Type of Linkage	Technologies	Product	Comments
1991_ cont'd	Yamanouchi Pharmaceutical	Mt. Sinai Medical Center	Collaborative research	Transgenic mouse		Collaboration to develop transgenic mouse model exhibiting Alzheimer's disease
?	Green Cross	University of California, San Francisco	Licensing agreement		MABs for cancer DX	MABs for cancer DX
?	Green Cross	University of California, Los Angeles	Collaborative research		Atrial peptide	Joint development agreement for atrial peptides
?	Sankyo	Washington University				

SOURCE: North Carolina Biotechnology Center *Actions Database*, BioScan, Japan Economic Institute Report, and other sources.

Appendix C

Workshop on U.S.-Japan Technology Linkages in Biotechnology: Agenda and Participants

Wednesday, June 12, 1991 - NAS Green Building Room 104
2001 Wisconsin Avenue, NW, Washington, D.C.

National Research Council's
Committee on Japan

Introductory Comments by Chairman
    HUBERT SCHOEMAKER, Centocor, Inc.

Future Global Technology and Industry Trends
    STEVE BURRILL, Ernst & Young (discussion leader)
    Comments by:
    Robert Easton, The Wilkerson Group
    Isao Karube, Tokyo University
    David MacCallum, Hambrecht & Quist

Break

Trends in Technology Linkages
    MARK DIBNER, North Carolina Biotechnology Center
        (discussion leader)
    Comments by:
    Fumio Kodama, Harvard University
    Roger Longman, Windhover Information, Inc.
    Joel Marcus, Brobeck, Phleger & Harrison

Senior Management Perspectives (Working Lunch)
    HUBERT SCHOEMAKER, Centocor, Inc. (discussion leader)
    Comments by:
    Yasuo Iriye, Otsuka America, Inc.

Roles for Universities and Government

> **JAMES WYNGAARDEN**, National Research Council, and
> **ROBERT YUAN**, University of Maryland (discussion leaders)
> Comments by:
> Marvin Cassman, National Institutes of Health
> Marvin Guthrie, Massachusetts General Hospital
> Susanne Huttner, University of California System
> Hideaki Yukawa, Mitsubishi Petrochemical Co.

Investment Issues

> **STELIOS PAPADOPOULOS**, PaineWebber (discussion leader)
> Comments by:
> Joseph Lacob, Kleiner Perkins Caufield & Byers
> Robert Riley, Bristol-Myers Squibb Co.
> Alan Walton, Oxford Partners

Concluding Discussion of "Big Picture Questions"

Closing Remarks by Chairman

Adjourn

Other Participants/Discussants

> Susan Clymer, NichiBei Bio, Inc.
> Michael Goldberg, American Society for Microbiology
> Joshua Lerner, Harvard University
> Rachel Levinson, Office of Science and Technology Policy
> Kathryn Lindquist, State of Maryland, Department of Economic and Employment Development
> Lesley Russell, U.S. House of Representatives Committee on Energy and Commerce
> Weijian Shan, University of Pennsylvania

Note: **BOLD** denotes members of the NRC biotechnology working group.